The Human Security Agenda

The Human Security Agenda

How Middle Power Leadership Defied US Hegemony

RONALD M. BEHRINGER

Continuum International Publishing Group
A Bloomsbury company

50 Bedford Square
London
WC1B 3DP

80 Maiden Lane
New York
NY 10038

www.continuumbooks.com

ISBN: 978-1-4411-3133-1 (hardcover)
978-1-4411-8299-9 (paperback)

Library of Congress Cataloging-in-Publication Data
A catalog record for this book is available from the Library of Congress.

Typeset by Deanta Global Publishing Services, Chennai, India
Printed in the United States of America

I dedicate this book, with love, to my mother, Helen Mina.

CONTENTS

Portions of this book were first published in Behringer, Ronald M. 2005. "Middle Power Leadership on the Human Security Agenda." *Cooperation and Conflict* 40(3): 305–42. *Cooperation and Conflict* is a Journal of the Nordic International Studies Association published by Sage Publications

ACKNOWLEDGMENTS

The ideas which I have expressed in this book have evolved considerably over my academic career, and I am grateful to several individuals who have helped me along the way to develop them. I began exploring the topic of middle power leadership on human security during my doctoral studies in political science at the University of Florida. I would like to thank my mentor and chair of my doctoral dissertation supervisory committee, Dr. Ido Oren from the Department of Political Science, for his advice and guidance throughout the dissertation research and writing process, as well as his suggestions for the title of this book. I am also grateful to the members of my supervisory committee—Dr. J. Samuel Barkin, Dr. M. Leann Brown, and Dr. Michael D. Martinez from the Department of Political Science, Dr. Terry L. McCoy from the Department of Political Science and the Department of Latin American Studies, and Dr. Paul J. Magnarella from the Department of Anthropology and the Levin College of Law—for providing me with helpful comments on the drafts of my chapters.

My gratitude goes out to Dr. Martinez, Dr. Barkin, and Dr. Goran S. Hyden for providing me with funding during their terms as Graduate Coordinator of the department. While writing my dissertation, I was awarded a Gibson Dissertation Fellowship by the College of Liberal Arts and Sciences at UF, and I would like to thank the College, Mr. Robert Gibson, and Mrs. Jean Gibson for their generosity. Furthermore, I am highly appreciative of the assistance which I received on countless occasions from the staff of the Department of Political Science, particularly Mrs. Debbie Wallen and Mrs. Suzanne Lawless-Yanchisin.

I am indebted to several individuals who assisted me with my research. The Honorable Mr. Jacques Saada—formerly the Canadian Minister Responsible for Democratic Reform, the Leader of the Government in the House of Commons, and the Member of Parliament for the electoral riding of Brossard-La Prairie, and currently the President and Chief Executive Officer of the Quebec Aerospace Association—helped me to arrange interviews in Ottawa. Mr. Tim Martin, the former Director of the Peacebuilding and Human Security Division of the Canadian Department of Foreign Affairs and International Trade (DFAIT), and the current Canadian Ambassador to Colombia, and Dr. Hélène Laverdière, the former Deputy Director of the Peacebuilding and Human Security Division of DFAIT, and current

Member of Parliament for the riding of Laurier—Sainte-Marie, were both gracious in taking the time to be interviewed.

I also appreciate the insights of three individuals who participated in e-mail interviews: Mr. Hans Hækkerup, the former Danish Minister for Defense (1993–2001), and the former Special Representative of the Secretary-General for Kosovo and Head of the United Nations Interim Administration Mission in Kosovo (UNMIK, 2001); Mrs. Mette Kjuel Nielsen, the former Head of the Department for Russia, CIS, OSCE, and the Balkans at the Danish Ministry of Foreign Affairs, former Danish Deputy Permanent Secretary of Defense (1998–2001), former Chair of the SHIRBRIG Steering Committee, and current Danish Ambassador to Serbia; and Dr. Peter Viggo Jakobsen, the former Head of the Department of Conflict and Security Studies at the Danish Institute for International Studies in Copenhagen, and current part-time lecturer at the University of Copenhagen's Department of Political Science. I would also like to thank Dr. Knud Erik Jørgensen from the Department of Political Science and Government at the University of Aarhus for encouraging me to publish a synopsis of my dissertation research in *Cooperation and Conflict* in 2005 when he was editor of the journal.

While I was teaching at Concordia University in Montreal, Mrs. Marie-Claire Antoine, the Acquisitions Editor for Politics and International Relations at Continuum Publishing, inquired about my research on human security. So began the process of publishing this book, and I would like to thank Marie-Claire, her Editorial Assistant at Continuum, Ally Jane Grossan, and the editorial team at Deanta Global Publishing Services, led by Subitha Nair, for their tremendous assistance. I would also like to thank Dr. W. Andy Knight, the Chair of the Department of Political Science at the University of Alberta, for his constructive criticism of my book proposal, which inspired me to develop my ideas further. Moreover, I am grateful to Dr. Peter Stoett, the former Chair of the Department of Political Science at Concordia University and Mrs. Jeannie Krumel, the Department Administrator, for granting me extended library privileges to assist me with preparing my book manuscript after my contract with Concordia had ended.

Finally, I could not have achieved my goals in life without the love and support of my mother, Helen Mina. She has been there for me at every step of the way, offering encouragement whenever I needed it most. I dedicate this book to her as an expression of my eternal gratitude.

LIST OF TABLES

LIST OF ABBREVIATIONS

ACDA	United States Arms Control and Disarmament Agency
AMA	American Medical Association
AMICC	American Non-Governmental Organizations Coalition for the International Criminal Court
AMIS	African Union Mission in the Sudan
AMISOM	African Union Mission in Somalia
APLs	Antipersonnel Landmines
APRODEH	*Asociacion pro Derechos Humanos* (Association for Human Rights)
ASIL	American Society of International Law
ASPA	American Service-Members' Protection Act
ATLs	Antitank Landmines
ATT	Arms Trade Treaty
AU	African Union
BIAs	Bilateral Immunity Agreements
CACM	Central American Common Market
CAR	Central African Republic
CARICOM	Caribbean Community and Common Market
CCPCJ	Commission on Crime Prevention and Criminal Justice
CCW	Convention on Certain Conventional Weapons
CD	Conference on Disarmament
CICC	NGO Coalition for an International Criminal Court
CIS	Commonwealth of Independent States
CSS	Critical Security Studies
CUNY	City University of New York

DFAIT	Canadian Department of Foreign Affairs and International Trade
DPKO	United Nations Department of Peacekeeping Operations
DRC	Democratic Republic of the Congo
ECOSOC	United Nations Economic and Social Council
ECOWAS	Economic Community of West African States
EU	European Union
FIDH	*Fédération International des Ligues des Droits de l'Homme* (International Federation of Leagues of Human Rights)
FORD	Friends of Rapid Deployment
FPLC	*Force Patriotique pour la Libération de Congo* (Patriotic Force for the Liberation of Congo)
GNP	Gross National Product
G77	Group of 77
HLP	High Level Panel on Threats, Challenges and Change
HRW	Human Rights Watch
HRW/PHR	Arms Project of Human Rights Watch and Physicians for Human Rights
IANSA	International Action Network on Small Arms
ICBL	International Campaign to Ban Landmines
ICC	International Criminal Court
ICID	International Commission of Inquiry on Darfur
ICISS	International Commission on Intervention and State Sovereignty
ICJ	International Court of Justice
ICRC	International Committee of the Red Cross
ICRtoP	International Coalition for the Responsibility to Protect
ICTR	United Nations International Criminal Tribunal for Rwanda
ICTY	United Nations International Criminal Tribunal for the Former Yugoslavia
IDPs	Internally Displaced People

IDRC	International Development Research Centre
ILC	International Law Commission
IMT	International Military Tribunal
IRIN	Integrated Regional Information Networks
JEM	Justice and Equality Movement
LMG	Like-Minded Group of Countries
LOI	Letter of Intent
LRA	Lord's Resistance Army
MINURCAT	United Nations Mission in the Central African Republic and Chad
MINURSO	United Nations Mission for the Referendum in Western Sahara
MINUSTAH	United Nations Stabilization Mission in Haiti
MLC	*Mouvement de libération du Congo* (Movement for the Liberation of Congo)
MONUSCO	United Nations Organization Stabilization Mission in the Democratic Republic of the Congo
MOU	Memorandum of Understanding
MOU/SB	Memorandum of Understanding on SHIRBRIG
MOU/SC	Memorandum of Understanding on the Steering Committee
NAM	Non-Aligned Movement
NATO	North Atlantic Treaty Organization
NGO	Nongovernmental Organization
NRA	National Rifle Association of America
OAS	Organization of American States
OAU	Organization of African Unity
ONUC	United Nations Operation in the Congo
ONUMOZ	United Nations Operation in Mozambique
ONUSAL	United Nations Observer Mission in El Salvador
OSCE	Organization for Security and Cooperation in Europe
P5	Permanent Five Members of the United Nations Security Council

PCIJ	Permanent Court of International Justice
PGA	Parliamentarians for Global Action
PLANELM	Planning Element
PoA	Program of Action
PTC	Pre-Trial Chamber
RDMHQ	Rapidly Deployable Mission Headquarters
R2P	Responsibility to Protect
RtoP	Responsibility to Protect
R2P-CS	Responsibility to Protect—Engaging Civil Society
SADC	Southern Africa Development Community
SALW	Small Arms and Light Weapons
SD	Self-Destruct
SDA	Self-Deactivation
SHIRBRIG	Multinational Standby High Readiness Brigade for United Nations Operations
SIPRI	Stockholm International Peace Research Institute
SLM/A	Sudan Liberation Movement/Army
TCI	Transfer Controls Initiative
UK	United Kingdom
UN	United Nations
UNAMA	United Nations Assistance Mission in Afghanistan
UNAMID	African Union/United Nations Hybrid operation in Darfur
UNAMIS	United Nations Advance Mission in Sudan
UNAVEM	United Nations Angola Verification Mission
UNDOF	United Nations Disengagement Observer Force
UNDP	United Nations Development Program
UNDPI	United Nations Department of Public Information
UNEF	United Nations Emergency Force
UNFICYP	United Nations Peacekeeping Force in Cyprus
UNGA	United Nations General Assembly
UNICEF	United Nations Children's Fund

UNIDIR	United Nations Institute for Disarmament Research
UNIFIL	United Nations Interim Force in Lebanon
UNISFA	United Nations Interim Security Force for Abyei
UNMD	United Nations Military Division
UNMEE	United Nations Mission in Ethiopia and Eritrea
UNMIK	United Nations Interim Administration Mission in Kosovo
UNMIL	United Nations Mission in Liberia
UNMIS	United Nations Mission in Sudan
UNMISS	United Nations Mission in the Republic of South Sudan
UNMIT	United Nations Integrated Mission in Timor-Leste
UNMOGIP	United Nations Military Observer Group in India and Pakistan
UNOCI	United Nations Operation in Côte d'Ivoire
UNODA	United Nations Office for Disarmament Affairs
UNOSOM	United Nations Operation in Somalia
UNSAS	United Nations Standby Arrangements System
UNSC	United Nations Security Council
UNSF	United Nations Security Force
UNSG	United Nations Secretary-General
UNTAG	United Nations Transition Assistance Group
UNTEA	United Nations Temporary Executive Authority
UNTSO	United Nations Truce Supervision Organization
UPC	*Union des Patriotes Congolais* (Union of Congolese Patriots)
US	United States
USAFSTC	United States Army Foreign Science and Technology Center
USCBL	United States Campaign to Ban Landmines
USDIA	United States Defense Intelligence Agency
VVAF	Vietnam Veterans of America Foundation
WFM	World Federalist Movement
WFM-IGP	World Federalist Movement-Institute for Global Policy

CHAPTER ONE

Introduction

Since the end of World War II, the realist paradigm has been the dominant approach for the study of international relations (Holsti 1995; Lynn-Jones 1999; Pettiford and Curley 1999). Realists conceptualize the dynamics of international security as a zero-sum struggle between nation-states in an anarchic international system (Morgenthau 1948). The great power states are considered to be the primary actors in the global system. International peace is maintained either through a balance of military power between the great powers (Waltz 1979) or through the actions of a globally hegemonic state that possesses preponderant military power and the will to exercise it (Gilpin 1981).

But the events of the post-Cold War era have demonstrated the limitations of the state-centric realist approach. Unlike earlier periods, when the main security concern for national governments was to prevent the outbreak of warfare between nation-states, there has been a proliferation of intrastate conflicts since the Cold War ended. National militaries, warlords, guerrillas, secessionist groups, and terrorist organizations have used deadly force against both military and civilian targets in pursuit of their objectives. The realist perspective, which prioritizes the security of nation-states from military threats, has been incapable of dealing with the post-Cold War conflicts, where the combatants are often nonstate actors, and the primary victims are civilian populations. The world became painfully aware of this fact on September 11, 2001, when members of the Al-Qaeda terrorist network attacked the World Trade Center and the Pentagon in the United States, killing around 3,000 people.

A particular deficiency of realism is its failure to recognize that smaller states may exercise leadership on security issues. The realist view of middle and small powers is summarized concisely in Robert Gilpin's argument that "both power and prestige function to ensure that the lesser states in the system will obey the commands of the dominant state or states" (Gilpin 1981, 30). But by concentrating overwhelmingly on the activities of great power states, realists have ignored the contributions that middle powers have made to global security. This book addresses the neglect of middle powers and demonstrates how some of these states have taken the initiative to make significant positive contributions to global security, even while facing powerful opposition from great powers.[1]

In 1996, the Canadian Minister for Foreign Affairs, Lloyd Axworthy, brought to the attention of the international community an alternative conceptualization of security, termed "human security," which emphasizes the security of people rather than the security of nation-states (Axworthy 1997). In recent years, human security issues have been promoted worldwide by "like-minded" middle powers such as Australia, Canada, Denmark, the Netherlands, and Norway. These states have launched human security initiatives in multilateral forums and have brokered global coalitions of the willing, which have worked diligently to achieve these initiatives. In short, middle powers have assumed leadership roles on human security issues, and they have played these roles successfully.

Formidable obstacles may lie in the path toward success, however. Negative reactions from other states should be expected whenever a human security proposal does not correspond to their national interests. One particular state, the hegemonic United States of America, stands alone above all other states due to its preponderant power, capabilities, and influence in the domain of global security. In this book, I explore the US policy on the human security issues, as well as the reaction of Washington to each middle-power-led initiative. I expect that if the US government should perceive a human security initiative as a threat to its core national interest—defined as the security of the American territory, institutions, and citizenry—the superpower would mount a fierce campaign to thwart the initiative. It can be safely asserted that the like-minded middle powers would never take any action that would willfully endanger the territory or population of the United States, a fellow democracy and close ally. But there is the possibility that an international institution (norm, law, or organization) that is established with the intention of furthering human security may simultaneously contradict American domestic institutions (laws and constitutional rights). In this scenario, I assume that Washington would defend the American institutions by countering the human security initiative.

The successful achievement of a human security objective may also depend on the middle powers' choice of diplomatic strategy, which I also explore in this book. Occasionally, the middle powers have opted to fulfill a human security initiative without the participation or assistance of any great powers. A possible benefit of this strategy is the maintenance of freedom from the great powers' influence, interference, and manipulation of an initiative to make it conform to the great powers' interests rather than serve those of humanity. A drawback, however, is that the middle powers may have limited capabilities and resources to mobilize on behalf of an initiative without contributions from the great powers.

For some human security issues, the middle power states have relied on the traditional decision-making by consensus which characterizes many multilateral organizations. The positive aspect of consensus decision-making is that it may generate broad support from the international community (including the great powers) for an initiative, thereby increasing the likelihood that the initiative will achieve its objectives. On the negative side, any state may single-handedly thwart an initiative on the grounds that it conflicts with the national interest of the dissenting state, no matter how incredulous its arguments are. Moreover, consensus decision-making often produces lowest common denominator agreements that have less substance and may not be satisfactory for fully ensuring human security.

In contrast, the middle powers have sometimes taken control of the negotiating process by utilizing "soft power" (Nye 1990; Sikkink 2002), "fast-track diplomacy" (Axworthy 2003; Lawson 1998), or a combination of both. Soft power involves the use of persuasion, rather than coercion, to convince actors to support an initiative. Rob McRae described soft power as "leadership through both example and exemplary actions, and through partnership" (McRae 2001a, 245). Fast-track diplomacy is a "take it or leave it" approach whereby the middle powers organize a coalition of like-minded actors—including states, international humanitarian organizations, and nongovernmental organizations (NGOs)—which have come to an agreement on a treaty or plan of action that is effective for addressing a particular human security problem. The coalition then uses the soft power of persuasion, through both state-led diplomacy and NGO-led advocacy, to convince as many holdout states as possible to accept the human security proposal within a brief period of time. Although fast-track agreements lack the universal support that characterizes accords reached through consensus, they have made tangible progress in resolving issues that affect human security.

The human security initiatives

In this book, I analyze five cases of human security initiatives on which middle powers played significant leadership roles. I will now summarize these initiatives, which are discussed in greater detail in subsequent chapters.

The SHIRBRIG peacekeeping initiative

The lack of a permanent force for rapid deployment in times of crisis has consistently hampered UN peacekeeping operations (Johansen 1998). To address the problem, the governments of Canada and the Netherlands organized the Friends of Rapid Deployment (FORD) coalition, with the aim of promoting the idea of a UN rapid response brigade among the major power states (Langille 2000). In December 1996, seven middle powers established the Multinational Standby High Readiness Brigade for United Nations Operations (SHIRBRIG). Headquartered in Denmark, SHIRBRIG was designed to deploy, at a short notice of only fifteen to thirty days, four to five thousand troops on peacekeeping missions lasting up to six months (SHIRBRIG 2001a, 2001b). SHIRBRIG was first deployed successfully as part of the UN Mission in Ethiopia and Eritrea (UNMEE) in 2000–2001, and was subsequently active in other African countries. At its height, around two dozen middle powers and small states participated in SHIRBRIG at four different levels of membership. SHIRBRIG was created and deployed without the assistance or participation of any great powers. Despite its accomplishments, a reduction in the global demand for traditional peace-keeping brigades forced SHIRBRIG to shut down permanently on June 30, 2009 (SHIRBRIG 2010a).

The Ottawa Process to ban antipersonnel landmines

The scourge of antipersonnel landmines (APLs) claims thousands of victims annually. In response to this crisis, the International Campaign to Ban Land-mines (ICBL), an umbrella group of NGOs, issued a call for a global ban on the use, production, stockpiling, and transfer of APLs (Williams and Goose 1998). The Canadian government decided to exercise leadership on the land-mines issue by co-hosting, together with the NGO Mines Action Canada, a conference in October 1996 on the issue of banning APLs (Lawson et al. 1998; Lenarcic 1998). At the conference, Canadian Foreign Minister Lloyd Axworthy invited the participants to work with Canada to negotiate and sign an APL ban treaty by December 1997, a mere fourteen months after the Ottawa conference. The Ottawa Process core group, consisting of like-minded states, engaged in fast-track negotiations on an APL ban treaty and networked with NGOs to cultivate global political support for the treaty. At the December 1997 landmines conference, also held in Ottawa, one hundred and twenty-two states signed the Convention on the Prohibition of the Use, Stockpiling, Production and Transfer of Anti-personnel Mines and on Their Destruction. The Ottawa Convention entered into force in March 1999 and had an immediate impact on curtailing the global production and transfer of APLs.

The International Criminal Court

The idea of creating a permanent institution that would pursue justice for crimes against humanity has been discussed for decades. In 1994, the Like-Minded Group of Countries (LMG) was formed, with the objective of campaigning for the establishment of an International Criminal Court (ICC) that would try cases of genocide, crimes against humanity, war crimes, and the crime of aggression (Pace and Schense 2001). The LMG developed a close working relationship with the NGO Coalition for an International Criminal Court (CICC), which professed its faith that the leadership and negotiating capabilities of the LMG would produce a strong ICC. Instead of settling for a lowest common denominator agreement, the LMG campaigned for an effective treaty, even though a few major powers expressed their opposition (Pace 1999; Robinson 2001). At the 1998 Rome Conference, skilled diplomacy by delegates from the LMG managed to persuade holdout governments to vote in favor of the Rome Statute, which established an independent ICC with the power to initiate its own investigations and prosecutions of crimes. The ICC came into effect on July 1, 2002, following the sixtieth ratification of the Rome Statute.

The campaign to regulate the legal trade in small arms and light weapons

Small arms and light weapons (SALW) have been responsible for the deaths of millions of people in intrastate conflicts in the post-Cold War era. But due to a preoccupation with major weapons systems during the Cold War, the global community failed to adopt international norms regarding the production, transfer, and possession of SALW (Renner 1999). It was not until 1993 that the SALW issue was placed on the international agenda, when Mali asked the UN for assistance with the uncontrolled proliferation of SALW in West Africa (Smaldone 1999). In August 1997, a UN Panel of Governmental Experts issued a report calling for the convening of an international conference on the illicit trade in SALW in all its aspects (Lozano 1999).

At the meetings of the Preparatory Committee prior to the conference, Canada submitted a working paper which recommended that an action plan on SALW should examine the relationship between the licit and the illicit aspects of the SALW problem, and that states should exercise maximum restraint on the legal manufacture and trade of SALW (UN Preparatory Committee for the UN Conference on the Illicit Trade in SALW 2000). The middle powers established a close working relationship with the NGO community, including the International Action Network on Small Arms (IANSA), and co-hosted a series of seminars and workshops that were designed to cultivate the global political will to address the problem of the SALW trade.

At the July 2001 UN Conference on the Illicit Trade in Small Arms and Light Weapons in All Its Aspects, the anti-SALW coalition pushed for a prohibition on the sale of SALW to nonstate actors and emphasized that the eradication of the illegal trade in SALW cannot be accomplished without first establishing stronger regulations on the legal trade (NGO Committee on Disarmament 2001c). Because negotiations at the conference were characterized by consensus decision-making, pressure from the US delegation was successful in ensuring that the final agreement would not include any references to the regulation of private gun ownership or a ban on SALW transfers to nonstate actors. The conference ended with the participants adopting a Program of Action (PoA) that is politically, but not legally, binding and addresses solely the illicit aspects of the SALW trade. Subsequent UN meetings have not dealt with the legal trade in SALW.

The responsibility to protect

The genocidal conflicts of the 1990s illustrated that the sovereignty principle may harm human security as much as it protects national security. In response, middle powers such as Canada and Australia took the lead in redefining sovereignty, emphasizing the responsibilities that states have toward their citizens. Chief of these is the responsibility to protect (R2P). The International Commission on Intervention and State Sovereignty (ICISS), established by the government of Canada and co-chaired by the former Australian foreign minister Gareth Evans, issued an influential report in 2001, which argued that R2P involves the responsibilities of prevention, reaction, and rebuilding (ICISS 2001a). The UN members endorsed the ICISS report and the notion of R2P at the 2005 World Summit, but modified the R2P concept significantly. The middle powers accepted the consensus decision-making of the World Summit negotiations in order to generate a R2P principle that, despite being watered down, would receive the widespread approval of the international community.

The structure of the book

This chapter presented an overview of my research. Chapter 2 provides the theoretical background to my study. The chapter discusses the evolution of security studies, the emergence of the human security agenda, the dynamics of "middlepowermanship," and the potential consequences of human security for the American national interest. I then engage in an analysis of middle power leadership on human security in subsequent chapters. Chapter 3 covers the SHIRBRIG initiative in rapidly deployable peacekeeping. In Chapter 4, I demonstrate how the Ottawa Process succeeded in achieving a ban on APLs. Chapter 5 looks at how middle power cooperation resulted in the creation of the ICC. In Chapter 6, I explain why the middle powers were unsuccessful in their attempt to regulate the legal trade in SALW. Chapter 7 examines the contributions of the middle powers to the development of the

R2P principle. Finally, Chapter 8 assesses the dynamics and the results of the human security initiatives. The future of human security, both as a theoretical approach and as a foreign policy objective, is evaluated as well.

Note

1 The arguments that I present in this book have been refined considerably since I analyzed middle power leadership on human security in Behringer 2005.

CHAPTER TWO

Human security and the middle powers: a theoretical background

It is rare that readers of international relations scholarship will encounter terms like "human security" or "middle powers." This is unfortunately due to the predominance of the realist approach to international relations and its traditional focus on securing nation-states (and especially great powers) from military attacks. I wish to amend this neglect in the literature. Human security is a new approach that provides both an alternative conceptualization of security and an agenda of important security issues that need to be resolved. The middle powers have played influential roles in redefining the global security agenda by embracing and promoting the notion of human security.

In this chapter, I challenge the contemporary relevance of traditional, myopic approaches to security. I do this by illustrating how the field of security studies has both "widened" in terms of the issues that are now considered to be potential threats to security and "deepened" with regards to the referent object of security. I also posit that actors other than great powers, namely middle power states, have exercised leadership in the domain of global security. I argue that the middle powers have played the roles of "securitizing actors" in emphasizing the need to protect human populations, and "norm entrepreneurs" in promoting the human security agenda. In order to accomplish my objectives, I will now turn to a review of the transforming nature of security studies and the emergence of the concept of human security. I will also explore how the middle powers have assumed leadership on the human security agenda, as well as the problems that the US national interest may pose for the fulfillment of human security initiatives.

The changing conceptualization of security

Widening and deepening in security studies

In their book on the evolution of international security studies, Barry Buzan and Lene Hansen (2009) emphasized that four questions have structured debates within security studies since the end of World War II. The first question is whether to focus exclusively on the state as the referent object of security? Newer approaches to security have drawn attention to the need to extend security to nonstate actors, such as human populations that live within or across state borders. This analytical shift can be termed a "deepening" of the referent object of security. The second question is whether to analyze internal security threats as well as external ones? It has become particularly evident in the post-Cold War era that threats to security may not only emanate from other countries, but may emerge from domestic actors as well. The third question is whether to include nonmilitary threats to security as well as military threats, or in other words, "widen" the range of potential security issues? The fourth question is whether to perceive security as connected to a context of threats? Is security always about the urgent need to protect oneself from an existential threat? Alternatively, would the concept of security change if there are corresponding changes in how the security threats and referent objects are defined?

Since the beginning of the Cold War era, international security studies has been dominated by the "traditionalists," who emphasize that priority should be given to securing the nation-state from military threats (Buzan and Hansen 2009; Morgan 2007; Walt 1991). The chief of the traditionalist approaches has been realism. Realist approaches have evolved over time, from classical realism emphasizing the power maximizing nature of states (Carr [1939] 1964; Morgenthau 1948), to structural realism (or neorealism) positing that states react to changes in the systemic distribution of power (Gilpin 1981; Waltz 1979), and further to neoclassical realism arguing that states' responses to international stimuli are conditioned by unit-level variables such as their domestic political-military cultures (Dueck 2005). Despite their differences, all realist approaches have been similarly state-centered in terms of their assumption that the nation-state is the primary actor of importance in international relations. Furthermore, they have been great power-centric in terms of their overwhelming preoccupation with the most powerful states in the international system. Realists are also united in their beliefs that governments prioritize national security over all other issues and that states should forever remain vigilant against possible military attacks from other countries.

The onset of the Cold War and systemic bipolarity in the late 1940s led to the establishment of Strategic Studies as the core of international security studies (Baylis et al. 1987; Buzan and Hansen 2009). Dominated by realists,

Strategic Studies focused on the global competition for power between the United States and the Soviet Union. Of particular interest to Strategic Studies were the pacifying dynamics of nuclear deterrence (Kissinger 1957; Schelling 1966), an issue which was easily amenable to quantification and statistical analysis, and thus en vogue with the behavioralist revolution in the American social sciences at the time.

The policies derived by Strategic Studies were challenged during the Cold War by numerous scholars who were concerned about the ramifications of nuclear proliferation and spiraling arms races, especially the potential for a World War III. The field of Peace Studies (Barash 2010; Buzan and Hansen 2009; Rogers 2007) was founded by mainly liberal and Marxist intellectuals with the intent of deriving an alternative approach that would displace the bellicose realism which had pervaded Strategic Studies. Peace Studies proponents argued that an enduring peace could not be based solely on the absence of war (a scenario they termed "negative peace"), but necessitated addressing injustices in the political, economic, and social realms which are the root causes of conflict (i.e. produce a situation of "positive peace").

With the dissolution of the Soviet Union and the end of the Cold War, the traditionalists lost the context of existential threat which had fuelled their research agenda (Buzan and Hansen 2009). There was growing dissatisfaction among scholars with security studies' narrow focus on military and nuclear threats to the state. This discord originated with the emergence of the economic and environmental agendas in global politics during the 1970s and 1980s, and intensified in the 1990s with the advent of identity issues and the increase in transnational crime (Buzan, Wæver, and de Wilde 1998). Moreover, the positivist and rationalist foundations of traditionalist research were being contested on epistemological grounds by the postpositivists, who encouraged reflectivist approaches instead (Lapid 1989; Wæver 1996). The resulting proliferation of paradigms started new debates in international security studies on the need for broader interpretations of the concept of security.

"Wideners" began to emerge in security studies calling for the analysis of both nonmilitary and military threats to security (Brown 2003; Buzan 1997; Buzan, Wæver, and de Wilde 1998; Mutimer 1999). In his landmark book, *People, States and Fear*, Barry Buzan (1983, 1991) presented "classical security complex theory," which posited that international security is relational and thus dependent on how actors relate to one another. Furthermore, it argued that since states were more likely to fear their neighbors rather than remote powers, the referent objects of security analysis should be regional subsystems. Buzan emphasized that security should be analyzed at the sectoral level in order to identify particular types of interaction. He described five sectors of security: military, political, economic, societal, and environmental. In an article in *Foreign Affairs*, Jessica Tuchman Mathews

(1989) supported the arguments of the wideners by stressing that the scope of security concerns has broadened to include resource, environmental, and demographic issues.

According to David Mutimer (2007), the term "Critical Security Studies" (CSS) was first coined at a May 1994 conference of security scholars in Toronto, with the intent of establishing an epistemic community of scholars from different theoretical persuasions who are united by their discontent with the traditionalist perspective on security.[1] The conference participants outlined a three-part agenda for CSS to fulfill. First, CSS would challenge the state's predominance as the referent object of security by encouraging a deepening of the referent object to other actors, such as nations or human populations. Second, CSS would support the wideners by exploring how multiple actors and phenomena, besides organized national militaries, could pose threats to security. Finally, contributors to CSS would adopt a postpositivist epistemology and would criticize the traditionalists for their claims of objectivity in studying international relations.

The Copenhagen School of security studies was launched in 1998 with the publication of *Security: A New Framework for Analysis* (Buzan, Wæver, and de Wilde 1998).[2] The authors built upon Buzan's (1983, 1991) work on classical security complex theory, but chose to extend the theory "to sectors other than the military-political and to actors other than states" (Buzan, Wæver, and de Wilde 1998, 16). Moreover, Barry Buzan, Ole Wæver, and Jaap de Wilde adopted a social constructivist approach in order to explain the process through which issues become securitized. "Securitization" means that a problem is "presented as an existential threat, requiring emergency measures and justifying actions outside the normal bounds of political procedure" (Buzan, Wæver, and de Wilde 1998, 23–4). The process of securitization necessitates a speech act from a "securitizing actor" who emphasizes that a particular phenomenon poses an imminent threat to the survival of a referent object (usually the state). This belief in the subjectivity of securitization contrasts with the traditionalists' conviction that "real" security threats exist in the world and that these threats can be studied objectively.[3]

Despite the emergence of Critical Security Studies and the Copenhagen School, the predominance of the traditionalists has not diminished in the post-Cold War era. Rather, the traditionalists have bolstered their centrality in security studies through a marriage between Strategic Studies and the more positivist and rationalist wings of Peace Studies, their epistemological mates (Buzan and Hansen 2009). The traditionalists have not been shy to castigate approaches that call for a widening of the security agenda. A particularly famous critique by Steven Walt (1991) went so far as to argue that an expansion in the scope of security studies to include issues unrelated to the threat or use of force would destroy the intellectual coherence of the discipline. But no matter what one's personal perspective is on the ramifications for the academic enterprise, there is no doubt that the new approaches have stimulated novel thinking about security.

Human security: the birth of a new approach

The advocates of a widening of the notion of security proved to be very influential on the United Nations Development Program (UNDP). In the 1994 edition of its annual *Human Development Report*, the UNDP elaborated the concept of human security for the first time (Hay 1999; UNDP 1994).[4] The report called for a reconceptualization of security, where the emphasis would shift from securing the nation-state from the threat of a nuclear attack to guaranteeing the security of human populations. This would entail not only a widening of the concept of security, but a deepening of it as well. The goal of the UNDP was to improve the quality of human life, by ensuring that people may live their lives in free and safe environments. This objective corresponded with the vision of the UN Secretary-General Boutros Boutros-Ghali, who claimed that threats to global security extend beyond the military sphere and include phenomena such as environmental degradation, drought, and disease (UN Secretary-General 1992 [hereafter UNSG]).

Fen Osler Hampson and his colleagues (2002) have described three different conceptions of human security that are currently in use, ranging from a minimalist to a maximalist definition. The first perspective is the "rights-based" approach to human security, which stresses the legal protection of human populations. The objective of this approach is to strengthen normative legal frameworks at the global and regional levels, while simultaneously institutionalizing human rights law and reforming legal and judicial systems within nation-states. International institutions are viewed as essential for the development of new human rights norms and for ensuring the harmonization of national standards and practices regarding human security.

The second conception of human security is the "safety of peoples" or "freedom from fear" perspective (Hampson et al. 2002; McRae 2001b). Proponents of this view emphasize the physical security of human populations. They make a clear distinction between combatants and noncombatants in war, and argue that the international community has a moral obligation to intervene in conflicts in order to protect noncombatants from endangerment. Proponents also emphasize that any intervention by the international community in a conflict situation should extend beyond the mere provision of emergency humanitarian relief to include substantial actions which address the underlying causes of the conflict in order to prevent it from reoccurring. Furthermore, the freedom from fear approach acknowledges that, in many countries, the state may be more of a security threat to its own citizens rather than a guarantor of their security, echoing the arguments made previously by wideners (Buzan 1983, 1991; Irwin 2001; Kolodziej 1992). Human security may be threatened by either strong or weak states. While some states have the institutional capacity to carry out brutal repression (e.g. China, Myanmar, and North Korea), other states are too weak to prevent subnational groups from engaging in bloody armed conflict (e.g. Afghanistan, the Democratic Republic of Congo, and Somalia).

The third, and broadest, view of human security is the "sustainable human development" or "freedom from want" approach (Hampson et al. 2002). The concept of sustainable development dates back to the 1987 Brundtland Report, which claimed that environmental stress may produce, as well as result from, political tensions and armed conflict (Hay 1999; Page and Redclift 2002; World Commission on Environment and Development 1987).[5] According to the sustainable human development perspective, the process of development should cater to the needs of people.[6] Proponents believe that in order to address the numerous types of human security threats, humanity must first resolve fundamental problems of inequality and social injustice. As Jessica Tuchman Mathews argued, human security should be "viewed as emerging from the conditions of daily life—food, shelter, employment, health, public safety—rather than flowing downward from a country's foreign relations and military strength" (Mathews 1997, 51). The 1994 *Human Development Report* described seven areas in which human security could be threatened: economic, food, health, environmental, personal, community, and political (UNDP 1994). Studies have examined the impact of environmental issues on human security, such as climate change (Stripple 2002), the availability of food resources (Sage 2002; Turner, Brownhill, and Kaara 2001), and water scarcity (Cocklin 2002). Some radical scholars have called for drastic reforms in global institutions, and a new North-South partnership, in order to guarantee human security (Mayor 1995; Ul Haq 1995).

Peter Stoett (1999) has also examined minimalist and maximalist definitions, but with regards to the concept of "security" as well as four critical threats to human security: state violence, environmental degradation, population displacement, and globalization. While each minimalist definition restricts itself to the legal definition of the term, its maximalist counterpart includes the broader sociopolitical ramifications of the concept. Theorizing about security on a spectrum, the minimalist conception would stress the protection of the nation-state from a military attack, a broadening of the concept would include the protection of various collectivities from a plethora of security threats, and the maximalist definition would address transnational and intrastate threats to human security. In short, the maximalist conceptualization features both a widening and a deepening of the notion of security. While Stoett acknowledged that concepts may "soften as they broaden," trading analytical precision for greater inclusion, he stressed that a maximalist conception of security is a more accurate description of the reality of the contemporary age (Stoett 1999, 120).

S. Neil MacFarlane and Yuen Foong Khong (2006) agreed with Stoett on the fundamental importance of conceptual scope. With regards to human security, they claimed that "too inclusive an approach will mean the loss of analytical traction; too narrow an approach will mean fixating on military-state security and an inability to comprehend the nature of many of the post-Cold War threats" (MacFarlane and Khong 2006, 245). In contrast to the more inclusionary freedom from fear and freedom from want perspectives on human security, MacFarlane and Khong argued that human security should be limited to ensuring "freedom from organized violence." By recognizing that sentient individuals who organize to kill are the primary security threats to human populations, effective strategies may be derived to counter these threats. In contrast, it may be difficult or impossible to implement preemptive strategies in the event of natural threats to human security, such as tsunamis or earthquakes, which the broader "freedom from fear" approach would include.

Taylor Owen (2008) criticized MacFarlane and Khong's conception of freedom from organized violence for omitting human security threats which lack a deliberate perpetrator, such as natural phenomena. Owen proposed instead a threshold definition of human security, which includes only threats that exceed a specific threshold of severity that is objectively measured. While the types of threats that are considered are unlimited, attention is placed only on the most serious threats which endanger lives.

On a personal note, I adhere to the freedom from fear perspective on human security. I feel that this definition is a middle ground between the more limited focus of MacFarlane and Khong's freedom from organized violence and the more expansive agenda encapsulated in freedom from want. Moreover, the freedom from fear approach has been adopted by the governments of middle powers such as Canada and Norway, and it has thus served as the foundation for tangible human security policies. The case studies presented in this book examine issues that affect the physical security of human populations, irrespective of whether a perpetrator exists who intends to harm a targeted population or a particular threshold of severity has been surpassed.

Any expansion in the security agenda would certainly not be appealing to the traditionalists. But Steven Lamy made an interesting argument that "the traditionalists are not opposed to discussions about human security. However, they embrace a more particularistic position that assumes that states have an obligation to provide human security for their own citizens, but no obligation to provide security for noncitizens" (Lamy 2002, 170). By "human security," it can be safely assumed that Lamy was referring to the traditionalists' emphasis on physical security (i.e. freedom from fear) rather than quality of life (i.e. freedom from want). A focus on national security entails the protection of both the country's territory and the population that lives within the state's borders, which certainly contributes to freedom from fear. But since the referent object of security for the traditionalists is the nation-state and not human beings, it is doubtful that traditionalists would make the plight of foreign populations a central concern of national security, unless their predicament poses a direct threat to the security of the state (e.g. through a massive influx of refugees that would overwhelm domestic institutions).

The concept of human security has been embraced by a few middle power states, which have incorporated this novel conceptualization of security into their foreign policies. But who are these "middle powers"? What roles do they play in international relations? How have they formulated and promoted the human security agenda? I will answer these questions in the following sections.

The middle powers and the human security agenda

The neglect of middle powers

The foreign policies of middle power states have received little attention from international relations scholars. The main reason for this is the long predominance of the realist paradigm, which has influenced the perceptions of scholars as to which issues and actors are worthy of study. In the realists' conceptualization of an anarchic, self-help, international system, the states which command preponderant military power are the most significant actors. In his seminal work, *Politics Among Nations*, the classical realist Hans Morgenthau (1948) provided only a brief mention of small states and ignored middle powers completely. Moreover, the only discussion of small states by structural realists (or neorealists) was in their debate on whether the international system induces small states to balance or to bandwagon when they are threatened by a more powerful country (Walt 1987; Waltz 1979).

Although various scholars have conducted research on middle power states, most of their works have been single country studies. Furthermore, middle power scholars have tended to focus on their country of nationality. Only a handful of studies have engaged in a comparative analysis of middle power foreign policies. One of the earliest efforts was made by the Western Middle Powers and Global Poverty Project, which produced four edited volumes on various aspects of middle power internationalism, including the international development and foreign aid policies of five Western middle powers (Helleiner 1990; Pratt 1989, 1990; Stokke 1989). Subsequent studies have examined the foreign policies of diverse middle powers such as Argentina, Australia, Brazil, Canada, India, Malaysia, Mexico, Norway, South Africa, Sweden, and Turkey (Cooper 1997b; Cooper, Higgott, and Nossal 1993; Hurrell et al. 2000).

Defining middle powers

One of the reasons why very little comparative work on middle power foreign policies has been conducted may be the lack of consensus on the criteria for classifying states as middle powers. Four definitions of middle powers have been used in the academic literature (Cooper, Higgott, and Nossal 1993). The first definition is *geographic*: middle powers are states that are situated between two great powers. Examples of middle powers under this definition include Poland, which is located between the traditional great power rivals Germany and Russia, and Turkey, which serves as the bridge between the Western culture of Europe and the Islamic culture of the Middle East. The second definition is *normative*: middle powers are perceived as more virtuous, trustworthy, and wiser than either great powers or small states, due to their tendency to rely on diplomatic mechanisms to resolve conflicts, rather than the use of force. The problem with this definition is that it can be easily challenged whenever the actions of the middle powers do not live up to their moral rhetoric.

The third, and most commonly used, definition is *positional*. Under this definition, middle powers are categorized in terms of their national power relative to great powers and small states. Scholars have reached different conclusions as to which states qualify as middle powers, based on alternative rankings of national power (Cox and Jacobson 1973; Handel 1981; Holbraad 1984; Wood 1990). These rankings have been calculated using assorted combinations of factors, including Gross National Product (GNP), GNP per capita, population, nuclear capability, and prestige. Despite its popularity, the positional definition has been criticized for generating "few, if any, common patterns of behavior as to how a particular group of middle or intermediate powers will behave internationally, because the variation in the types of states involved, the categories of power that they possess, and the arenas within which they operate are all so various" (Hurrell 2000, 1).

The geographic, normative, and positional definitions of middle powers have been supplanted in recent years by a fourth definition that is based on *behavior*. According to this definition, middle power states are characterized by their performance of middle power diplomacy (Neack 2000), or "middlepowermanship," which Andrew Cooper, Richard Higgott, and Kim Richard Nossal described as "[the] tendency to pursue multilateral solutions to international problems, [the] tendency to embrace compromise positions in international disputes, and [the] tendency to embrace notions of 'good international citizenship' to guide . . . diplomacy" (Cooper, Higgott, and Nossal 1993, 19).[7] The "like-minded" middle powers, such as Canada, Denmark, the Netherlands, Norway, and Sweden, also share a "humane internationalist" outlook in their foreign policies. Humane internationalism features "an acceptance that the citizens and governments of the industrialized world have ethical responsibilities towards those beyond their borders who are suffering severely and who live in abject poverty" (Pratt 1990, 5). I adopt a behavioral definition of middle powers in this book.

Middlepowermanship in action

Gareth Evans, the former Australian Minister for Foreign Affairs and Trade (1988–96), argued that middle powers perform "niche diplomacy," which involves "concentrating resources in specific areas best able to generate returns worth having, rather than trying to cover the field" (Cooper, Higgott, and Nossal 1993, 25). Some middle powers have become renowned for their technical expertise in niche areas, such as Canada in the domain of peacekeeping (Hayes 1997) and Sweden on the issue of foreign aid (Elgström 1992). Middle power states have exercised leadership by acting as "catalysts" in launching diplomatic initiatives, "facilitators" in setting agendas and building coalitions of support, and "managers" in aiding the establishment of regulatory institutions (Cooper 1997a).[8] Multilateral institutions, such as the United Nations and regional organizations, have served as forums for effective middlepowermanship (Cooper 1992; Henrikson 1997; Keating 1993).

The soft power resources possessed by middle powers, such as their capacity to inform and persuade through the use of multilateral diplomacy and communications technologies, are becoming essential for competent leadership in a post-Cold War world that features greater interdependence and transnational cooperation (Nye 1990; Sikkink 2002). In a December 2003 interview, Tim Martin, the Director of the Peacebuilding and Human Security Division of the Canadian Department of Foreign Affairs and International Trade (DFAIT), mentioned a few possible conditions for effective middle power leadership: "policy analysis capability, credibility, predictability, communications, consistency, and diplomatic capacity."[9]

A study by Andrew Cooper, Richard Higgott, and Kim Richard Nossal (1993) claimed that the middle powers have opportunities to exercise leadership on international economic and social issues, due to a diminishing capacity and will of the United States to lead in these areas. Robert Cox (1989) stressed that the middle powers play a supporting role in the maintenance of global order and the development of international organizations and law, whether a hegemon exists or not. In a later work, Higgott suggested that middle powers "with the technical and entrepreneurial skills to build coalitions and advance and manage initiatives must show leadership when it is not forthcoming from the major actors" (Higgott 1997, 33). Gary Goertz argued that middle-power-led coalitions are frequently successful in achieving their objectives even when faced with great power opposition because "the major impetus [for an initiative] comes from smaller countries with larger ones coming in once they see that [the initiative] cannot be prevented" (Goertz 2003, 179).

But Cooper, Higgott, and Nossal did emphasize that since security concerns predominate in Washington, the middle powers have far less scope for international leadership in this domain. Instead, the middle powers tend to be supportive followers of great power leadership on the global security agenda. Cooper, Higgott, and Nossal stressed that the passive role of the follower is as important as the active role of the leader because "multilateralism and coalition-building can only work if there are more states willing [to] agree to join coalitions as followers than there are states seeking to play a leadership role" (Cooper, Higgott, and Nossal 1993, 118).

Thus, even some scholars of middlepowermanship have joined the realists in depicting the middle powers as passive followers of great power leadership in the realm of international security. I challenge this depiction of middle power states, however, by illustrating how certain middle powers have played leadership roles on human security initiatives. These middle powers have initiated campaigns to resolve particular issues that affect human security and have organized coalitions of like-minded states, international humanitarian organizations, and NGOs that have been effective mobilizing forces for the fulfillment of their human security objectives. The middle powers have used their influence to promote the establishment of new human security norms in international relations, to which I will now turn.

Creating international norms of human security

The traditional norms that have governed international relations and protected state sovereignty for centuries are possibly being supplemented by new norms protecting human security. Cristina Badescu and Thomas Weiss defined norms as "shared understandings and values that shape the preferences and identities of state and nonstate actors that legitimize behavior, either explicitly or implicitly" (Badescu and Weiss 2010, 358). Norms delineate what is considered to be appropriate behavior by actors (Finnemore and Sikkink 1998; Florini 1996). Ann Florini stressed that "the most important characteristic of a norm is that it is considered a *legitimate* behavioral claim . . . Norms are obeyed not because they are enforced, but because they are seen as legitimate" (Florini 1996, 364–5 [emphasis in the original]).

Florini (1996) presented an evolutionary model of norm change. She argued that norms are similar to genes in that both carry instructions. While genes instruct a biological organism in the case of organic evolution, norms instruct an individual, state, or other referent object in the case of normative evolution. International norms evolve over time due to the process of selection. Florini emphasized that three factors are necessary conditions, but not sufficient conditions on their own, for determining whether or not a particular norm will be selected from a norm pool and will spread throughout the international system. First, a norm needs to gain *prominence* in the norm pool vis-à-vis competitor norms. This usually occurs either through the efforts of a "norm entrepreneur" (an actor who attempts to change the behavior of others) to promote the norm or through states emulating the behavior of a prestigious state which adheres to the norm. Second, the degree to which a norm will be perceived as legitimate depends on its *coherence* with other prevalent norms. A norm that has a high degree of consistency with existing international norms and laws will be more resistant to change than a norm that is less coherent. Third, the selection of a norm is affected by *external environmental conditions*, including the distribution of power between states.

According to Florini, norm reproduction may occur either vertically (by future generations being socialized to adhere to a certain norm) or horizontally (by actors emulating the behavior of others who follow a norm). While vertical reproduction will lead to the continuation of a norm, horizontal reproduction will produce a change in norms. Florini described three conditions that facilitate horizontal reproduction: a drastic turnover in decision-makers (such as through a revolution), the undeniable failure of existing norms, and the rise of new issue areas in which no normative framework has yet been established.

Martha Finnemore and Kathryn Sikkink (1998) provided a theory of norm evolution, which they called the "norm life cycle." During the first stage of "norm emergence," norm entrepreneurs attempt to persuade a critical mass of states (the "norm leaders") to adopt a new norm. Once the norm leaders have accepted the norm, they will start convincing other states to follow suit. As soon as a "tipping point" is reached (usually when at least one third of the states have adopted the norm), the second stage of "norm cascade" is launched, whereby more states rapidly jump on the bandwagon of support for the norm. During the third stage of "internalization," states will develop considerable respect for the new norm and will automatically modify their behavior to comply with it.

The middle powers have been playing the roles of norm entrepreneurs in promoting the adoption and institutionalization of human security norms by the international community (Knight 2001). This corresponds with Andrew Cooper's (1997a) depiction of how the middle powers exercise leadership as catalysts, facilitators, and managers. The governments of the middle powers have identified areas of deficiency in the international normative and legal framework that have severe repercussions for humanity worldwide. They have selected new norms that would protect human security, and they have launched diplomatic initiatives to promote these norms globally. The middle powers have forged huge coalitions, featuring both state and nonstate actors, to rally support for the human security norms, and they have aided the institutionalization of these norms. But as Gareth Evans indicated, the middle powers have to be selective in terms of which issues they prioritize as part of their niche diplomacy (Cooper, Higgott, and Nossal 1993). The extent of middle power activism has depended considerably on the amount of resources that each state has been able and willing to devote for particular human security issues. Thus, the human security agenda reflects the interests and capabilities of the middle powers that set the agenda.

Establishing the human security agenda

The human security agenda received global attention in 1996, with the Canadian government's appointment of Lloyd Axworthy as Minister for Foreign Affairs (Hay 1999). In his address to the fifty-first United Nations General Assembly in September 1996, Axworthy explained that human security includes "security against economic privation, an acceptable quality of life, and a guarantee of fundamental human rights" (Axworthy 1997, 184). Moreover, Axworthy added that human security "acknowledges that sustained economic development, human rights and fundamental freedoms, the rule of law, good governance, sustainable development, and social equity are as important to global peace as arms control and disarmament" (Axworthy 1997, 184).

Axworthy clearly enunciated a maximalist definition of human security, which stressed the need to ensure both freedom from fear and freedom from want. Despite his noble intentions, the middle powers have tended to make more modest commitments to human security. Even the Canadian government of Prime Minister Jean Chrétien, in which Lloyd Axworthy served as foreign minister for four years, chose to base its human security policy solely on freedom from fear (Canada 2000, 2002), and placed many of the issues that Axworthy viewed as critical on the back burner. Rosalind Irwin indicated that "much of the human security agenda in Canada and elsewhere has downplayed economic and social concerns, focusing instead on the protection of individual security from violence" (Irwin 2001, 6). Norway has joined Canada at the forefront of those embracing a freedom from fear perspective (MacFarlane and Khong 2006). In contrast, Japan has been a vocal proponent of a freedom from want approach and has stressed that the achievement of peace and security depends on sustainable human development. S. Neil MacFarlane and Yuen Foong Khong (2006) indicated why Japan's human security policy has been based on a different perspective:

> Japan's embrace of a primarily developmentalist understanding of human security reflected . . . the country's discomfort with the seemingly interventionist thrust of the evolving discourse on human security. The link between protection and intervention was unacceptable to a number of key states in the region (notably China) with whom Japan had complex relations. Japan's unhappiness was exacerbated by the emerging consensus of the International Commission on Intervention and State Sovereignty on 'the responsibility to protect' (MacFarlane and Khong 2006, 159).[10]

In May 1998, Canada and Norway signed the Lysøen Declaration, a bilateral partnership for action on human security (Canada 2000, 2002; Lamy 2002; Small 2001). The Lysøen Declaration specified a framework for consultation and cooperation between the two middle powers. It also outlined an agenda of nine human security issues for Canada and Norway to prioritize. These issues were landmines, the International Criminal Court, human rights, international humanitarian law, gender dimensions in peace-building, small arms proliferation, children in armed conflict (including the problem of child soldiers), child labor, and Arctic and northern cooperation.

The bilateral Lysøen Declaration was expanded into a multilateral arrangement with the creation of the Human Security Network in September 1998, through which thirteen middle powers and small states work together on human security issues. As of January 2010, the members of the Human Security Network were Austria, Canada, Chile, Costa Rica, Greece, Ireland, Jordan, Mali, Norway, Slovenia, Switzerland, and Thailand, while South Africa was participating as an observer.[11] Tim Martin described the Human Security Network as "an advocacy and policy dialogue or mechanism."[12] According to Steven Lamy, the Human Security Network perceives itself as "a potential lobby group within existing international institutions" (Lamy 2002, 172). The objective of the network is to build coalitions of like-minded states within multilateral institutions in order to promote issues related to human security. But while the middle powers have preferred multilateralism, they have not always conformed to the norms of multilateral decision-making, as can be seen in the following section.

Consensus decision-making versus fast-track diplomacy

Multilateral diplomacy has increasingly featured consensus decision-making, which G. R. Berridge defined as "the attempt to achieve the agreement of all the participants in a multilateral conference without the need for a vote and its inevitable divisiveness" (Berridge 2005, 168). Consensus differs from unanimity, in that consensus can be reached as long as no delegates voice opposition to a measure, whereas unanimity requires that all delegates support an initiative. Jean-Robert Leguey-Feilleux described consensus decision-making as "a procedure that is expeditious and quite popular . . . a low-key procedure that can facilitate some decisions" (Leguey-Feilleux 2009, 225).

In a conference or international organization that features consensus decision-making, delegates will usually communicate informally to the officers in charge that they are not opposed to a certain proposal that is being negotiated. During the final meeting on the proposal, the chair will express their understanding that consensus has been reached (based on the informal poll) and ask if any delegate disagrees. If a delegate opposes consensus decision-making (i.e. they did not express themselves in favor of the proposal earlier during the informal poll, or they changed their mind since then), they solely need to request that a vote be held, which will then occur according to established voting rules. But if there are no objections, the chair will announce that the proposition is approved and will not record any vote. The proposal will thus be adopted through consensus (Leguey-Feilleux 2009).

The growing trend toward consensus decision-making has been due to the fear, particularly in wealthy states that provide most of the funding to international organizations, that majority voting would produce radical and irresponsible decisions spearheaded by the less developed states, which have formed the majority of the memberships in most international organizations since the 1960s (Berridge 2005). Consensus decision-making became popular following its effectual use during the Third UN Conference on the Law of the Sea (1973–82). Most of the decision-making in the United Nations system, including in organizations with weighted-voting systems such as the International Monetary Fund and the World Bank, tends to be based on consensus rather than voting.

But the problem with consensus decision-making is that it often produces watered-down, lowest common denominator agreements that have wide acceptance, but less substance. The middle powers have come to the realization that if they wish to make tangible progress in achieving human security, it may be more advantageous at times to take control of the negotiating process by utilizing "fast-track diplomacy" (Axworthy 2003; Lawson 1998). In this "take it or leave it" approach, the middle powers will organize a coalition of the "like-minded," whose membership may include certain states, international organizations, and NGOs which have reached an agreement on a convention or plan of action that is effectual for addressing a specific human security problem. The coalition will then employ the soft power of persuasion, through a combination of state-led diplomacy and NGO-led advocacy, to encourage as many holdout states as possible to endorse the human security proposal within a short period of time. Although a fast-track agreement lacks the universality of a pact reached through consensus, it can make substantial progress in resolving an issue that imperils human security.

Furthermore, fast-track diplomacy may facilitate the achievement of a human security initiative in a situation where a minority of states—and perhaps even a sole recalcitrant state—oppose the initiative because it is in conflict with their national interests. It is important to remember that all national governments evaluate courses of action in terms of their national interest as defined by the government. A government that does not believe in emerging human security norms will either refrain from supporting them or will attempt to thwart them if they should be in conflict with its national interest. Fast-track diplomacy is a method through which the humanitarian interest, including human security, may be served while sidestepping the hurdles erected by narrowly defined and parochial national interests.

Based on their track record, I believe that the middle powers are significant actors on the stage of global politics, and that they have exercised considerable leadership in the domain of human security. Applying the terminology of the Copenhagen School (Buzan, Wæver, and de Wilde 1998), the middle powers have been playing the roles of securitizing actors. They have used speech acts to publicize, and middlepowermanship to address, the existential threats that certain phenomena pose to human populations, the referent objects the middle powers have securitized. But leadership does not always produce desired results. Whether or not a human security campaign is ultimately successful may depend to a certain degree on how the United States, the global hegemon in the contemporary era, reacts to the initiative. It should be expected that Washington would evaluate any human security initiative in terms of its relevance or consequences for its national interest, and would react accordingly. But what does the American "national interest" entail? Moreover, is there a primary national interest that is prioritized over other interests? This shall be explored in the next section.

The core national interest of the United States

In his seminal work, *The Idea of National Interest*, Charles A. Beard emphasized that "national interest—its maintenance, advancement, and defense by the various means and instrumentalities of political power—is the prime consideration of diplomacy" (Beard 1934, 21). The essential components of the national interest were described by Scott Burchill (2005). First, the members of a society share numerous critical interests vis-à-vis outsiders, with survival or national security usually given the highest priority. Second, certain national interests are permanent and are above partisan politics. That is, these interests will not be altered whenever there is a change of government. Joseph Frankel (1970) termed these "objective" national interests, which may be discovered through scientific enquiry and which may include strategic goals such as acquiring a geopolitical advantage and maintaining control over material resources. Third, the national interest is defined by the national government on behalf of the citizens of the state.

Peter Trubowitz provided an alternative view that "the very definition of the national interest is a product of politics" (Trubowitz 1998, 4). The concept of a single national interest is a fictional construct. Instead, the national interest is determined by societal actors who are able to wield the necessary power within the political system to fulfill their policy preferences. Adherents to this perspective include Liberal approaches which posit that the state can fall captive to particular groups who define the national interest in terms of their own special interests (Frieden 1988; Milner 1987; Snyder 1991; Uslaner 1998), and Marxist approaches which argue that the bourgeoisie controlling the state attempts to disguise its own class interests as the national interest (Cox 1981; Marx and Engels [1848] 1986). Frankel (1970) called these "subjective" national interests that vary as governments alter, which may include ideology and class identity.

Realists emphasize that a state's primary interest is its national security, and that the means by which a state can maintain its security is through maximizing its power. According to the eminent realist scholar Hans Morgenthau, "the national interest of a peace-loving nation can only be defined in terms of national security, and national security must be defined as integrity of the national territory and of its institutions" (Morgenthau 1978, 553). Stephen Krasner agreed that the core objective of a state is the protection of its territorial and political integrity, but made a bolder claim that for a hegemonic state, "political and territorial integrity is completely secure" (Krasner 1978, 35). Karl Von Vorys stressed that the vital interest of any state is its national existence:

> National interest as a standard rests on one fundamental assumption: that the nation-state is for its citizens the focus of orientation or, in our specific case, that for Americans the secure existence of the United States is an altogether nonnegotiable value. It is never in the U.S. national interest to be conquered or subjugated by a foreign power. It is never in our national interest to delegate or transfer voluntarily to any foreign country or international organization our right to control our internal affairs and our right to conduct our foreign policy according to our own purposes. (Von Vorys 1990, 19)

Since the like-minded middle powers share a democratic political culture with the United States—and most of these countries participate in a military alliance with the superpower as well—it is certain that they would never engage in any actions that would threaten the territorial integrity of the United States or cause physical harm to the American populace. But a middle-power-led initiative could still pose a challenge to the US core national interest. Donald Nuechterlein claimed that "the fundamental national interest of the United States is the defense and well-being of its citizens, its territory, *and the US constitutional system*" (Nuechterlein 2001, 15 [emphasis added]). There is a possibility that the international institutions (norms, laws, or organizations) which a particular human security initiative seeks to establish may conflict with the domestic institutions of the United States, including the constitutional rights of American citizens. Since the preservation of the integrity of the US Constitution is a primary national interest of the United States, Washington would probably take action to defeat any initiative that infringes on constitutional rights.

The most important institution in the United States is the constitutional system, because it protects the principles of liberty, political equality, and self-government on which the country was founded (Spalding 2002). John Hall and Charles Lindholm argued that Americans have an "aura of sacredness" about the Declaration of Independence, the Constitution, and the Bill of Rights (Hall and Lindholm 1999, 92). In the words of John McElroy, the Constitution "declared the people's rights as the sovereign power, which the government was forbidden to infringe" (McElroy 1999, 166).

But post-Cold War developments in the field of international law may pose a threat to the sanctity of American constitutional rights. Lee Casey and David Rivkin (2003) indicated that the "traditional international law" (or the "law of nations"), which prohibits any violation of the sovereignty principle, is being displaced by the "new international law," which asserts that the domestic practices of states must conform with the emergent global norms of conduct. Casey and Rivkin are wary of the consequences of the new international law for the United States:

> As a philosophical matter, any attack upon the principle of sovereignty threatens the very foundation of American democracy. Sovereignty is the necessary predicate of self-government. . . . Any limitation on sovereignty as an organizing principle . . . is an abdication of the right of the citizens of the United States to be governed solely in accordance with their Constitution, and by individuals whom they have elected and who are ultimately accountable to them. To the extent that international law allows supranational, or extra-national, institutions to determine whether the actions of the United States are lawful, ultimate authority will no longer be vested in the American people, but in these institutions. (Casey and Rivkin 2003, 6–7)

In short, it is possible that any human security initiative which aims to establish new international institutions will also infringe on the sovereignty principle. These infringements are justified, as states should not be permitted to use sovereignty as a shield while they engage in genocide and human rights violations within their borders. But the new international laws may also clash with domestic laws that protect the rights of citizens in democratic states. As the global hegemon, the United States is likely to resist any human security initiative that conflicts with its core national interest. Thus, we should expect that Washington will actively oppose any human security proposal that it perceives as a potential threat to the constitutional rights of American citizens.

Conclusion

In this chapter, I have examined the widening and deepening of security studies, the development of the concept of human security, the leadership roles played by the middle powers as securitizing actors vis-à-vis human populations and norm entrepreneurs on the human security agenda, the diplomatic techniques the middle powers have used to fulfill this agenda, and the probability that the United States will resist human security initiatives that it perceives as threatening to its principal national interest. The lion's share of this book will analyze the dynamics of middlepowermanship on five key human security issues, as well as the reaction of the hegemonic United States to each initiative. I shall begin in the next chapter by exploring how the middle powers established SHIRBRIG.

Notes

1 The ideas generated at the conference were later published in Krause and Williams 1997. Buzan and Hansen (2009) claimed that the broad and inclusionary definition of Critical Security Studies expressed at the Toronto conference, and promoted by Krause and Williams, never gained popularity and was displaced by a narrower definition of CSS that was articulated by the Aberystwyth (Welsh) school (Booth 2005). This more limited conceptualization of CSS excluded constructivism, poststructuralism, feminism, and the Copenhagen School. See Mutimer 2007.
2 The proponents of the Copenhagen School have distinguished it from Critical Security Studies. While CSS believes that change is possible in international relations due to the socially constituted nature of phenomena, the Copenhagen School argues that "the socially constituted is often sedimented as structure and becomes so relatively stable as practice that one must do analysis also on the basis that it continues . . . This leads us to a stronger emphasis on collectivities and on understanding thresholds that trigger securitization in order to avoid them" (Buzan, Wæver, and de Wilde 1998, 35). Furthermore, rather than make normative claims about the benefits of security as CSS does, the Copenhagen School restricts itself to understanding the process of securitization.
3 The Copenhagen School has been criticized for its epistemological incongruity (Mutimer 2007). While the concept of securitization is based on a postpositivist epistemology, the sectoral analysis of security is rooted in positivism. This use of objective analysis, together with a reluctance to issue calls to emancipatory action, has led the Copenhagen School to be excluded from CSS by more radical approaches, such as the Aberystwyth School. See Booth 2005.
4 According to Rosalind Irwin (2001), the concept of human security has its roots in the adoption of the UN Charter in 1945 and the Universal Declaration of Human Rights in 1948. For a thorough discussion of how notions of security have evolved over the millennia, see MacFarlane and Khong 2006.
5 The official title of the Brundtland Report was *Our Common Future*. The report was produced by the World Commission on Environment and Development, also known as the Brundtland Commission, named after its Chairperson, the former Norwegian Prime Minister Gro Harlem Brundtland.
6 Robin Hay added that the sustainable human development perspective "branched off to include sustainable human security, which also places people squarely at the center of its concerns" (Hay 1999, 218).
7 The term "middlepowermanship" appears to have first been used by John Holmes (1965) and Paul Painchaud (1965) in papers given at the Third Annual Banff Conference on World Development in August 1965.
8 Andrew Cooper's conception of "catalysts" is similar to John Kingdon's "policy entrepreneurs," whom the latter defined as "advocates who are willing to invest their resources—time, energy, reputation, money—to promote a position in return for anticipated future gain in the form of material, purposive, or solidary benefits" (Kingdon 1984, 188).

9 Personal interview of Mr. Tim Martin, the Director of the Peacebuilding and Human Security Division of the Canadian DFAIT, Ottawa, Ontario, Canada, December 2, 2003.

10 Japan has been active in the area of sustainable human development, creating the UN Trust Fund for Human Security in 1998 and supporting UN Secretary-General Kofi Annan in his establishment of the Commission on Human Security in 2001. The Commission was co-chaired by Amartya Sen, the Nobel laureate in Economics, and Sadako Ogata, the Japanese former UN High Commissioner for Refugees. See MacFarlane and Khong 2006.

11 One founding member, the Netherlands, left the Human Security Network in 2007.

12 Personal interview of Mr. Tim Martin, December 2, 2003.

CHAPTER THREE

The SHIRBRIG initiative in rapidly deployable peacekeeping

Peacekeeping has been one of the most heralded activities of the United Nations over the past half-century. Thousands of military and civilian personnel have been deployed by the UN in dozens of peacekeeping missions around the world. Peacekeeping operations have experienced both considerable successes and dismal failures, and have been followed by an endless stream of criticism and calls for reform. But the considerable role that peacekeeping missions have played in the preservation of human security is indisputable. The presence of the "blue helmets" in a zone of conflict frequently deters armed militias from both attacking military targets and slaughtering civilian populations. The postconflict desire for retribution may fuel the murderous rampages of ex-soldiers and street thugs, and the innocent become victims on the mere basis of ascribed differences, such as ethnicity, language, or religion. The proliferation of bloody intrastate conflicts in the post-Cold War era necessitates that the UN be capable of intervening quickly enough to prevent and redress violations of human security.

It has been recognized that in order to make UN peacekeeping operations more effective, a standby peacekeeping force that is rapidly deployable to areas of conflict should be established (Johansen 1998; UNSG 1992; Urquhart 1993). The middle powers took the initiative to address this issue of human security. In 1996, seven middle power states created the Multinational Standby High Readiness Brigade for United Nations Operations

(SHIRBRIG). Four years later, SHIRBRIG was deployed with great success as part of the UN Mission in Ethiopia and Eritrea (UNMEE).

This chapter examines how the middle powers exercised leadership in establishing SHIRBRIG. I begin with an explanation of the practice of peacekeeping. The debate on rapid deployment is discussed, followed by the creation of the UN Standby Arrangements System (UNSAS). I then analyze how the middle powers cooperated in forming SHIRBRIG and review the accomplishments of the brigade. Finally, I describe the reaction of the United States to the SHIRBRIG initiative.

The practice of peacekeeping

"Peacekeeping" has been defined in various ways by scholars. According to John Hillen, "peacekeeping is a military technique for controlling armed conflict and promoting conflict resolution" (Hillen 1998, 79). Paul Diehl provided a more detailed definition:

> Peacekeeping is . . . the imposition of neutral and lightly armed interposition forces following a cessation of armed hostilities, and with the permission of the state on whose territory these forces are deployed, in order to discourage a renewal of military conflict and promote an environment under which the underlying dispute can be resolved. (Diehl 1993, 13)

Although Claus Heje insisted that peacekeeping did not originate with the United Nations, and that the UN is not the sole international organization to engage in peacekeeping, his definition centered on UN peacekeeping operations:

> As the United Nations practice has evolved over the years, a peacekeeping operation has come to be defined as an operation involving military personnel, but without enforcement powers, undertaken by the United Nations to help maintain or restore international peace and security in areas of conflict. (Heje 1998, 2)

Steven Ratner's definition of peacekeeping was also restricted to United Nations missions:

> Peacekeeping is . . . the stationing of UN military personnel, with the consent of warring states, to monitor ceasefires and dissuade violations through interposition between armies, pending a political settlement. (Ratner 1995, 17)

Peacekeepers are responsible for both observing the peace, through monitoring cease-fires and reporting breaches, as well as keeping the peace, by interposing between belligerents and creating buffer zones of disengagement (Weiss, Forsythe, and Coate 1997). Although there are no concrete rules for UN peacekeeping operations, there are a few principles which guide the practice of peacekeeping (Heje 1998). First, the parties in conflict must provide their consent to the establishment and continued operation of a peacekeeping mission, as well as to its mandate, composition, and commanding officer. Second, a peacekeeping mission may not violate the sovereignty of a host state. Third, peacekeepers must remain impartial between the conflicting parties. Fourth, peacekeepers may not use force except as a last resort in self-defense. Fifth, all peacekeeping operations must receive a clear mandate from the UN Security Council. Sixth, each peacekeeping mission requires a multinational deployment under the operational command and control of the UN Secretary-General. Finally, the participants in a peacekeeping mission must be willing to provide troops, financial aid, and logistical support to the operation.

The authorization for UN peacekeeping missions comes from Chapter VI (Articles 33 through 38) of the Charter of the United Nations, which is concerned with the pacific settlement of disputes that are likely to endanger peace (Pirnie and Simons 1996). Under Chapter VI, the UN Security Council sanctions the use of lethal force in self-defense while fulfilling a UN mandate. But the UN Charter does not mention peacekeeping specifically, hence, UN Secretary-General Dag Hammarskjöld coined the term "Chapter six-and-a-half," which refers to peacekeeping operations as an expansion of Chapter VI (Weiss, Forsythe, and Coate 1997).

"Traditional peacekeeping" operations have usually involved the deployment of multinational military contingents, consisting of a few thousand lightly armed soldiers grouped in light infantry battalions, in buffer areas between belligerent parties (Hillen 1998). The majority of infantry battalions have been donated either by middle power states such as Canada, Finland, Ireland, Norway, and Sweden or by countries that are neutral to a conflict, as Ghana and Nepal have done (Huldt 1995).

According to the United Nations, the first UN peacekeeping mission was the United Nations Truce Supervision Organization (UNTSO), which was deployed in the Middle East in June 1948 following the first Arab-Israeli War and is still in operation more than sixty years later (Goulding 1993). But some scholars have argued that UNTSO should not be labeled as a peacekeeping operation, since only military observers, who were either unarmed or equipped solely with side arms, participated in the mission (Hillen 1998; Macdonald 1998). They claimed that UN peacekeeping began with the United Nations Emergency Force (UNEF I), which was deployed from November 1956 until June 1967 in the Suez Canal, the Sinai, and the

Gaza Strip to maintain peace between the Arab states and Israel. UNEF I, much heralded as a model peacekeeping operation which established the principles and standards for future missions, was the brainchild of an exemplary diplomat from a middle power state. John Hillen suggested that the Canadian Secretary of External Affairs Lester Pearson's idea of sending an armed UN emergency force to interpose between the belligerent parties in Egypt "was a compromise of sorts between a small unarmed observation force such as that in Palestine [UNTSO] and a large and coercive collective security action such as the UN Command in Korea" (Hillen 1998, 82). In the words of Patricia Fortier,

> The 1956 Suez model of peacekeeping for which Lester B. Pearson won a Nobel Prize was built on the concept of the neutral observation of agreed behavior, and military resources in support of diplomatic agreements. The aim was to stall conflict between nation-states, thereby avoiding direct involvement by Great Powers. The basic premise for all peacekeeping operations was that states were the forces to be dealt with, and that the insertion of a recognized global authority backed by neutral military was acceptable to them. (Fortier 2001, 42)

Until the end of the Cold War, UN peace support operations engaged in the traditional peacekeeping described above. But with the proliferation of intrastate conflicts in the 1990s, it became evident that war criminals were using the inviolability of the sovereignty principle as a shield from justice while they carried out tortures, rapes, and murders of civilians. Traditional UN peacekeeping operations cannot protect populations if their governments do not consent to the deployment of peacekeepers on their soil.

In response, the United Nations broadened the mandates of its peace support operations in the post-Cold War era. New doctrines were formulated and implemented. "Peace enforcement" has been defined as "the deployment by the UN's political organs of military personnel to engage in non-consensual action, which may include the use of force, to restore international peace and security" (Ratner 1995, 19). Peace enforcement operations fall under Chapter VII (Articles 39 through 51) of the UN Charter (Pirnie and Simons 1996).[1] Chapter VII missions have the authorization of the Security Council to use lethal force beyond self-defense in order to accomplish a mandate. Article 39 describes three situations where the Security Council may authorize peace enforcement: a threat to the peace, a breach of the peace, and an act of aggression (Asada 1995). Peace enforcement necessitates that the UN surrender its impartiality, in order to authorize coercive measures that would compel a state to halt its violations of international law (Daniel and Hayes with Oudraat 1999). During the Cold War, the United Nations engaged in peace enforcement on a couple of occasions: the UN intervention in Korea (1950–53) and the UN Operation in the Congo (ONUC, 1960–64). Since the Cold War ended, the UN has enforced the peace in numerous cases, either by sanctioning the use of force by member states, such as during the 1991 Persian Gulf War against Iraq, or by authorizing an existing peacekeeping mission to employ force, as the UN Operation in Somalia (UNOSOM) was permitted to do in November 1992 (Ratner 1995).

Former UN Secretary-General Boutros Boutros-Ghali emphasized that following the resolution of a conflict, peacekeeping must be complemented by "peace-building," which he defined as "comprehensive efforts to identify and support structures which will tend to consolidate peace and advance a sense of confidence and well-being among people" (UN Secretary-General 1992, para. 55). Boutros-Ghali also described the various actions that constitute his vision of peace-building:

> These may include disarming the previously warring parties and the restoration of order, the custody and possible destruction of weapons, repatriating refugees, advisory and training support for security personnel, monitoring elections, advancing efforts to protect human rights, reforming or strengthening governmental institutions and promoting formal and informal processes of political participation. (UNSG 1992, para. 55)

Peace-building addresses the criticism of traditional peacekeeping operations that they have been "essentially passive and were meant only to defend the status quo by interposing a buffer between hostile parties" (Asada 1995, 36). Some peacekeeping missions, such as the UN Peacekeeping Force in Cyprus (UNFICYP) and the UN Interim Force in Lebanon (UNIFIL), have been deployed for decades without demonstrating any potential for resolving the underlying causes of the conflicts. The newer peace-building missions, however, have deployed peacekeeping forces only after a final settlement of the dispute has been reached. In contrast to the traditional peacekeeping of "Chapter VI ½," peace-building missions have been described by Masahiko Asada as "operations in 'Chapter VI ¼,' as they are approaching Chapter VI by encompassing more conflict resolution/prevention elements" (Asada 1995, 38). The UN Transition Assistance Group (UNTAG), which was established in April 1989 to ensure the successful implementation of the United Nations' plan for Namibian independence, is considered to be the first UN peace-building operation.[2] Other examples of peace-building missions include the UN Angola Verification Mission II (UNAVEM II), the UN Observer Mission in El Salvador (ONUSAL), and the UN Operation in Mozambique (ONUMOZ). In short, peace-building is concerned with consolidating the peace, even through nation-building, in order to prevent a recurrence of violence.

As illustrated in Table 3.1, there are sixteen UN peace support operations deployed as of September 2011. Five missions that were deployed during the Cold War to engage in traditional peacekeeping are still in the field. Six missions are currently authorized to enforce the peace when necessary. The United Nations is involved in peace-building activities in ten of

Table 3.1 United Nations peace support operations (September 2011)

Mission	Date first established	Location	Current personnel	Mandate
UN Truce Supervision Organization (UNTSO)	May 1948	Egypt, Israel, Jordan, Lebanon, Syria	148 MILOBS[a], 94 ICP[b], 120 LCS[c]	Traditional peacekeeping
UN Military Observer Group in India and Pakistan (UNMOGIP)	January 1949	Jammu and Kashmir	38 MILOBS, 25 ICP, 52 LCS	Traditional peacekeeping
UN Peacekeeping Force in Cyprus (UNFICYP)	March 1964	Cyprus	857 troops, 63 CIVPOL[d], 40 ICP, 112 LCS	Traditional peacekeeping
UN Disengagement Observer Force (UNDOF)	June 1974	Syrian Golan Heights	1,036 troops, 43 ICP, 102 LCS	Traditional peacekeeping
UN Interim Force in Lebanon (UNIFIL)	March 1978	Southern Lebanon	11,746 troops, 351 ICP, 656 LCS	Traditional peacekeeping
UN Mission for the Referendum in Western Sahara (MINURSO)	April 1991	Western Sahara	27 troops, 197 MILOBS, 4 CIVPOL, 98 ICP, 162 LCS, 18 UNV[e]	Peacekeeping, peace-building
UN Interim Administration Mission in Kosovo (UNMIK)	June 1999	Kosovo	8 military staff, 146 ICP, 236 LCS, 28 UNV[f]	Peace-building, nation-building

UN Assistance Mission in Afghanistan (UNAMA)	March 2002	Afghanistan	More than 1,600 civilian staff (around 80% LCS)	Peace-building
UN Mission in Liberia (UNMIL)	September 2003	Liberia	7,782 troops, 130 MILOBS, 1,288 CIVPOL, 471 ICP, 997 LCS, 233 UNV	Peacekeeping, peace-building
UN Operation in Côte d'Ivoire (UNOCI)	April 2004	Côte d'Ivoire	8,974 troops, 193 MILOBS, 1,276 CIVPOL, 397 ICP, 743 LCS, 216 UNV	Peacekeeping, peace enforcement, peace-building
UN Stabilization Mission in Haiti (MINUSTAH)	June 2004	Haiti	8,728 troops, 3,524 CIVPOL, 564 ICP, 1,338 LCS, 221 UNV	Peacekeeping, peace enforcement, peace-building
UN Integrated Mission in Timor-Leste (UNMIT)	August 2006	Timor-Leste	33 MILOBS, 1,194 CIVPOL, 396 ICP, 890 LCS, 193 UNV	Peace-building
African Union/United Nations Hybrid operation in Darfur (UNAMID)	July 2007	Darfur, Sudan	17,759 troops, 311 MILOBS, 4,526 CIVPOL, 1,136 ICP, 2,834 LCS, 480 UNV	Peacekeeping, peace enforcement, peace-building
UN Organization Stabilization Mission in the Democratic Republic of the Congo (MONUSCO)	July 2010	Democratic Republic of the Congo and the subregion	17,010 troops, 746 MILOBS, 1,241 CIVPOL, 983 ICP, 2,828 LCS, 580 UNV	Peacekeeping, peace enforcement, peace-building

(Continued)

Table 3.1 (Continued)

Mission	Date first established	Location	Current personnel	Mandate
UN Interim Security Force for Abyei (UNISFA)	June 2011	Abyei area, Sudan	Maximum of 4,200 troops, 50 CIVPOL, and appropriate civilian support[g]	Peacekeeping, peace enforcement
UN Mission in the Republic of South Sudan (UNMISS)	July 2011	South Sudan	Maximum of 7,000 troops, 900 CIVPOL, and appropriate civilian support[g]	Peacekeeping, peace enforcement, peace-building, nation-building

Source: United Nations. 2011b. "United Nations Peacekeeping." <http://www.un.org/en/peacekeeping/> (Accessed 12 September 2011).

Source for data on UNMIK: United Nations Mission in Kosovo. 2011. "About UNMIK." <http://www.unmikonline.org/Pages/about.aspx> (Accessed 12 September 2011).

Source for data on UNAMA: United Nations Assistance Mission in Afghanistan. 2011. "UNAMA: United Nations Assistance Mission in Afghanistan." <http://unama.unmissions.org/Default.aspx?tabid=1741>(Accessed 12 September 2011).

Notes: The personnel levels for troops, military observers, civilian police, and UN volunteers are as of 31 July 2011, while the personnel levels for international civilian personnel and local civilian staff are as of 30 June 2011, unless stated otherwise.

[a]MILOBS = Military Observers.

[b]ICP = International Civilian Personnel.

[c]LCS = Local Civilian Staff.

[d]CIVPOL = Civilian Police.

[e]UNV = United Nations Volunteers.

[f]The personnel levels for UNMIK are as of 31 December 2010.

[g]The personnel levels for UNISFA and UNMISS are the maximum authorized, rather than the number deployed.

its missions. The UN Interim Administration Mission in Kosovo (UNMIK) and the UN Mission in the Republic of South Sudan (UNMISS) are also engaged in nation-building, while the UN Assistance Mission in Afghanistan (UNAMA) is a political mission directed and supported by the UN Department of Peacekeeping Operations (DPKO).

The debate on rapid deployment

The lack of a permanent brigade that is capable of responding rapidly to crisis situations has been a major problem for UN peacekeeping operations (Johansen 1998). In his June 1992 report, titled *An Agenda for Peace*, UN Secretary-General Boutros Boutros-Ghali recommended that the United Nations should create a rapid reaction force, which, under Article 40 of the UN Charter (provisional measures), could be used in peace enforcement operations (UNSG 1992). The publication of *An Agenda for Peace* sparked a considerable amount of debate on the need for a UN peacekeeping brigade. Peter Langille (2000) distinguished between the "practitioners," who called for the strengthening of the existing peacekeeping arrangements, and the "visionaries," who promoted the idea of establishing a UN standing brigade or rapid response capability.

The dialogue between Sir Brian Urquhart and François Heisbourg reflects the debate between the visionaries and the practitioners (Urquhart and Heisbourg 1998). Urquhart, the former UN Under-Secretary-General for Special Political Affairs (1974–86), reiterated the argument, first enunciated by UN Secretary-General Trygve Lie in 1954, for the creation of a standing, volunteer, rapid response brigade of around ten thousand troops under the control of the UN Secretariat (Lie 1954). In Urquhart's view, a rapid reaction brigade would not take the place of traditional peacekeeping forces nor of peace enforcement missions, but would be deployed as "an immediate practical response to conflict or potential conflict at a point where quite a small effort might achieve disproportionately large results" (Urquhart and Heisbourg 1998, 192). The rapid reaction force would only remain in the field during the acute phase of the crisis and would be replaced by a regular peacekeeping mission as soon as the UN could organize one. The tasks of a rapid response brigade would include establishing a UN presence in the zone of crisis; preventing an escalation in violence; assisting and monitoring a cease-fire; providing the emergency framework for UN efforts at conflict resolution; securing a base and infrastructure for a subsequent UN peacekeeping operation; ensuring safe areas for persons and groups who are threatened by the conflict; securing humanitarian relief operations; and furnishing the Security Council with primary assessments of the conflict. Furthermore, a rapid reaction force would need sufficient logistical support and arms to ensure its own mobility and security, but would never be assigned military objectives under Chapter VII of the UN Charter.

In response to Brian Urquhart, François Heisbourg concurred on the desirability of a UN rapid response capability, but questioned its feasibility from a military standpoint (Urquhart and Heisbourg 1998). The UN force would be expected to perform, in an effective manner, a wider array of tasks than national militaries do. The nature of rapid response activities, such as securing safe havens, makes it impossible to maintain the United Nations' impartiality with regards to a conflict. The UN would also have to deal with the consequences of pursuing coercive and partial objectives. In addition, rapid deployment would depend on the prompt adoption of resolutions by the Security Council, a dubious assumption given the Council's track record. Since a rapid response capability does not correspond neatly to either Chapter VI or Chapter VI½, establishing such a capability would necessitate redefinitions of UN peace support operations. Moreover, Heisbourg estimated that the annual cost of maintaining a rapid reaction brigade consisting of ten thousand personnel would exceed $300 million.[3] This figure does not include logistical expenses, or the costs related to the multinational participation in the brigade. Finally, while Heisbourg admitted that creating a rapid reaction group is better than maintaining the status quo, he argued that improving standby peacekeeping arrangements may be a more realistic option.

A 1993 article by Brian Urquhart on the need for a UN volunteer military force has also received considerable criticism (Urquhart 1993; Urquhart et al. 1997). Lee Hamilton (D-Indiana), the former Chairman of the US House of Representatives Committee on Foreign Affairs, suggested that while Urquhart's proposal merits a lengthy consideration, the present UN peacekeeping system should be reformed in the meantime. Hamilton insisted that the United States could assist the UN in ensuring collective security by negotiating an Article 43 special agreement to provide military units to the UN on short notice, contributing surplus material to the UN stock of peacekeeping equipment, sharing information, and paying dues on time. Another critique came from Gareth Evans, the former Foreign Minister of Australia, who questioned the capability of a five thousand troop brigade to make a significant impact when much larger interventions have failed.[4] Moreover, Field-Marshal Lord Carver warned that a rapid response force may serve merely as reinforcement for the weaker side in a conflict, thereby prolonging the fighting and discouraging the parties from deriving an enduring political settlement. Finally, Stanley Hoffmann cautioned that the composition of a rapid reaction brigade "would have to be carefully balanced so as not to allow for any suspicion of great power predominance or of manipulation by an interested regional power" (Urquhart et al. 1997, 149).

There is strong empirical evidence to support the argument that the UN should develop a rapid response capability. Robert Johansen (1998) mentioned two cases where the failure to respond promptly had disastrous consequences. First, if the UN had possessed a rapid reaction brigade in July 1990, it could have been deployed at the Kuwaiti border before Iraqi forces invaded. A preventive deployment could have possibly deterred Saddam Hussein from attacking Kuwait, by indicating the resolve of the international community to stand up to acts of aggression. Second, if a UN rapid response force had been available when the fighting began in Rwanda in April 1994, the genocidal massacres of thousands of people may have been prevented by the swift creation of safe havens for refugees.

The United Nations Standby Arrangements System (UNSAS)

In 1993, UN Secretary-General Boutros Boutros-Ghali called for the creation of a system of standby arrangements which would furnish the personnel and equipment needed for the rapid deployment of peacekeeping operations (Langille 2000). The United Nations Standby Arrangements System (UNSAS) is based on each member state making a conditional commitment, under Article 43, to provide a specified amount of resources for UN peacekeeping operations within a predetermined response time. These resources include soldiers, police, and civilian personnel, as well as equipment and specialized services. Participating states maintain their pledged resources on standby mode and provide the necessary training and preparations in accordance with UN guidelines. The national contingents are expected to deploy within thirty days of a Security Council mandate for a traditional peacekeeping mission and within ninety days for a complex mission (United Nations Military Division 2001 [hereafter UNMD]).

According to Peter Langille (2000), UNSAS serves four objectives. First, it furnishes the UN with information about the capabilities of member states to contribute to peacekeeping missions at particular points in time. Second, it aids the planning, preparation, and training for peacekeeping operations. Third, the UN is provided with a set of options in the event that certain member states choose not to participate in a mission. Finally, standby arrangements may encourage member states to maintain their commitments to peacekeeping operations.

Secretary-General Boutros-Ghali emphasized that not only would UNSAS enable the UN to respond with greater speed and cost-effectiveness, it would assist the member states in planning and budgeting for their UN peacekeeping contributions (UNSG 1994a). In 1994, Boutros-Ghali established a Standby Arrangements Management Team within the DPKO that would clarify the UN requirements in peacekeeping missions, negotiate with those member states who wished to participate, derive readiness standards, create a peacekeeping resource database, help with mission planning, and formulate new procedures for determining the reimbursement of the member states' equipment that is used in peacekeeping operations (Langille 2000).

By June 1994, twenty-one member states had already pledged a total of thirty thousand personnel as standby resources, and another twenty-seven states had commitments pending (UNSG 1994b). Seventy-seven states had joined UNSAS by 2003 (UNMD 2003b). In July 2002, the new "Rapid Deployment Level" in UNSAS came into effect, which is a level of commitment where states pledge resources that can be deployed to a UN mission within thirty to ninety days of the adoption of a Security Council mandate (UN Military Adviser 2002). But only two states, Jordan and Uruguay, agreed to this level of commitment, and between them they placed a mere six units on standby for UN peacekeeping operations. Although several UN missions have been planned with the support of UNSAS (UNMD 2003a), the system does not yet provide a rapid deployment capability for UN peacekeeping operations.

The middle powers take the initiative

Middle power proposals for a brigade

The middle power states assumed leadership on the issue of enhancing the rapid response capability of UN peacekeeping missions. In 1994, the Dutch government conducted a study on the prospects for establishing a standing, rapid-response UN brigade, and an international conference was held to discuss the study's results (Langille 2000). The Dutch government then released *A UN Rapid Deployment Brigade: 'A Preliminary Study'* in April 1995 (The Netherlands 1995). The report argued that, instead of focusing its attention on strengthening UNSAS, the United Nations should set up a permanent, rapidly deployable force. The recommendations for a standing brigade were similar to those made by Brian Urquhart (1993) and Robert Johansen (1998). The brigade would engage in either preventive action, emergency relief during a humanitarian crisis, or interim peacekeeping during the period between a Security Council decision to deploy a peacekeeping mission and the arrival of the mission in the field.

The report estimated that a brigade numbering five thousand personnel would cost around $300 million annually and that $500–50 million would be needed for the initial purchasing of equipment (The Netherlands 1995). In order to reduce the expenses of equipment procurement, basing, and transportation, the report suggested that the brigade should be "adopted" by an international organization, such as the North Atlantic Treaty Organization (NATO), or by one or more member states. Although the Dutch proposal received some support, most states rejected the idea of a standing UN brigade, and refused to pay even the modest expenses suggested by the Dutch. In the words of Peter Langille, "it was clear . . . that only a less binding, less ambitious arrangement would be acceptable, at least for the immediate future" (Langille 2000, 223).

In September 1995, the Canadian government presented the UN with a report titled *Towards a Rapid Reaction Capability for the United Nations* (Canada 1995). The study provided twenty-one recommendations for enhancing the United Nations' peacekeeping mechanisms in the short to medium term and five suggestions for peacekeeping innovations in the long term. The Canadian report adviced that the UN should develop a rapid reaction capacity, and highlighted the requirements for such a capability: "an early warning mechanism, an effective decision-making process, reliable transportation and infrastructure, logistical support, sufficient finances, and well-trained and equipped personnel" (Langille 2000, 223). The study identified the need for a permanent, multinational, rapid response headquarters, consisting of thirty to fifty personnel, who would manage the prompt deployment of peacekeepers. It was suggested that the UN should create multinational groups and give each group the responsibility for a different function related to peace support operations. In order to do so, the UN would need to persuade member states to place specialized "vanguard units" on standby in their home countries, and link them, through UNSAS, to the rapid response headquarters. In contrast to the Dutch study, the Canadian initiative supported the improvement of the existing peacekeeping system, including UNSAS. The intention of the Canadian proposal was to launch a cooperative, pragmatic, low-cost effort at reforming UN peacekeeping.

A third proposal came from the government of Denmark, which announced in January 1995 that it would seek the support of other states in creating a multinational working group to study the feasibility of establishing a UN rapid reaction force (Denmark 1995a). The proposal was the brainchild of Hans Hækkerup, the Danish Minister of Defense, whose leadership would prove to be instrumental for the creation of a UN rapid deployment capability.[5] As Hækkerup explained in a 2004 interview:

> After discussions with then UN Under-Secretary-General for Peacekeeping Operations Kofi Annan in 1993, I started to develop the SHIRBRIG concept, and convinced my Canadian and Dutch colleagues that SHIRBRIG was not in contradiction with their proposals, but rather a stepping stone in that direction.[6]

Between May and August 1995, Denmark hosted four international seminars in which twelve middle power states and the DPKO participated (Langille 2000).[7] The *Report by the Working Group on a Multinational UN Standby Forces High Readiness Brigade*, which was issued in August 1995, argued that it was possible for a group of member states to combine their contributions to UNSAS in order to create a Multinational Standby High Readiness Brigade for United Nations Operations (SHIRBRIG), that would be deployable, at a short notice of only fifteen to thirty days, on peacekeeping operations for up to 180 days (Denmark 1995b; SHIRBRIG 2001a). The brigade would be responsible for carrying out peacekeeping and humanitarian tasks under Chapter VI of the UN Charter, and would be required to protect UN agencies and personnel, as well as NGOs, in the field.

The Danish-led multinational study called on participants to adopt standardized training and equipment, as well as hold joint exercises, in order to facilitate the deployment of the brigade (Denmark 1995b; SHIRBRIG 2001a). Since participation in peacekeeping operations would be voluntary for member states, the multinational report emphasized that a "brigade pool" should be established, consisting of a surplus of units beyond the force requirement for the brigade when it is deployed. The brigade pool would ensure that the brigade could be deployed with sufficient resources even if some member states abstained from participating in a mission. The report also stressed that the brigade would need to be self-sufficient for a period of sixty days. Furthermore, the participating states would need to cooperate in providing logistics to the brigade, including forward supply bases, because the brigade would have to be capable of operating in an environment where support from the host state is lacking, or where the infrastructure is either in poor condition or nonexistent.

The foreign ministers of the Netherlands, Canada, and Denmark viewed these three studies as making a mutual contribution to the development of a UN rapid response capability (Langille 2000). During the commemoration of the United Nations' fiftieth anniversary, Canadian Foreign Minister André Ouellet and Dutch Foreign Minister Hans Van Mierlo organized a meeting of ministers from nine middle powers and small states in order to rally support for a UN rapid reaction force.[8] Canada and the Netherlands then set up an informal group called the "Friends of Rapid Deployment" (FORD), which was co-chaired by the Canadian and Dutch permanent representatives to the UN. The objective of FORD was to promote the idea of a UN rapid deployment brigade among the major powers, and it used the Canadian report as the starting point for discussion. By the autumn of 1996, FORD had expanded to include twenty-six states, but only three members, Brazil, Germany, and Japan, were major powers.[9] FORD had also begun cooperating with the UN Secretariat and the DPKO. UN Secretary-General Boutros Boutros-Ghali had recently expressed in his 1995 *Supplement to An Agenda for Peace* that he had "come to the conclusion that the United Nations does need to give serious thought to the idea of a rapid reaction force" (UNSG 1995, para. 44).

FORD's original focus was on generating support for the proposals of the 1995 Canadian study, namely, the creation of a rapidly deployable headquarters, the improvement of UNSAS, and the elaboration of the concept of "vanguard units" (Langille 2000). But since the SHIRBRIG model described in the Danish-led multinational report also incorporated elements of the vanguard concept, FORD soon switched its emphasis toward promoting the Danish initiative. Peter Viggo Jakobsen, the Head of the Department of Conflict and Security Studies at the Danish Institute for International Studies, indicated that the Dutch proposal for a UN standing army would have been preferable to a standby brigade, but it was not a politically feasible plan like SHIRBRIG.[10] FORD assisted in the enhancement of UNSAS and helped the DPKO set up a Rapidly Deployable Mission Headquarters (RDMHQ), consisting of military and civilian personnel, as proposed in the Canadian study. The rapid deployment initiative of FORD encountered some roadblocks, however. Several nonaligned states expressed their displeasure that they were not included in FORD and accused the latter of being an illegitimate group. The nonaligned states also raised the question of equitable representation in both SHIRBRIG and the RDMHQ, and were particularly vocal at the annual spring meetings of the UN Special Committee on Peacekeeping Operations (also known as the Committee of 34). As Hans Hækkerup described:

> SHIRBRIG was also met with considerable skepticism in the UN system and among some of the major third world troop contributors like Pakistan. This was very UN-like, reflected in a long discussion whether SHIRBRIG could call itself a UN Standby Force. The problem was subsequently solved by deleting UN from the official name![11]

SHIRBRIG becomes a reality

Despite the criticism from the nonaligned states, the like-minded middle powers carried on full steam with the SHIRBRIG initiative. The process of setting up SHIRBRIG involved the signing of four documents (SHIRBRIG 2001a, 2001b, 2001c). On December 15, 1996, Austria, Canada, Denmark, the Netherlands, Norway, Poland, and Sweden signed the first document, the Letter of Intent (LOI) to cooperate on establishing a framework for SHIRBRIG that would be based on the recommendations of the Danish-led multinational study. By signing the LOI, a state became an "Observer Nation" in the Steering Committee, the executive body of SHIRBRIG. The second document that was signed was the Memorandum of Understanding on the Steering Committee (MOU/SC). The states which signed this document were permitted to participate in the development of SHIRBRIG policies in the Steering Committee.

The third document was the Memorandum of Understanding on SHIRBRIG (MOU/SB). In signing this document, a state agreed to commit troops to the brigade pool. The final document was the Memorandum of Understanding on the Planning Element (MOU/PLANELM). The planning element (PLANELM), located in Høvelte Kaserne, Denmark, was a permanent staff of thirteen military officers drawn from ten states. Each state which had signed the MOU/PLANELM agreed to station one or two staff officers in the PLANELM. In its predeployment stage, the PLANELM was responsible for deriving standard operating procedures for SHIRBRIG, working on concepts of operations, and organizing and conducting joint exercises. During deployment, the PLANELM was to be expanded to include eighty-five officers and noncommissioned officers, and would serve as the hub of the brigade headquarters staff numbering 150 personnel. At full deployment, SHIRBRIG could mobilize four to five thousand soldiers, who would be assigned to a headquarters unit with communication facilities, infantry battalions, reconnaissance units, medical units, engineering units, logistical support, helicopters, or military police (SHIRBRIG 2001c).

By November 2008, twenty-three states were participating in SHIRBRIG at four different levels of membership (SHIRBRIG 2010a, 2010b). Most of these countries could have been classified as middle powers, though some of them were small states. Ten countries had signed all four SHIRBRIG documents, and were considered to be full members: Austria, Canada, Denmark, Italy, the Netherlands, Norway, Poland, Romania, Spain, and Sweden.[12] Finland, Lithuania, and Slovenia had signed all documents except the MOU/PLANELM. Ireland had signed the LOI and the MOU/SC. Portugal had signed the LOI solely, and thus served as an Observer Nation in the Steering Committee. Although they had not signed the LOI, Chile, Croatia, the Czech Republic, Egypt, Jordan, Latvia, and Senegal were also designated as Observer Nations because they had expressed interest in the SHIRBRIG initiative, and possibly would have signed one or more SHIRBRIG documents in the future.[13]

The Steering Committee determined six criteria for participation in SHIRBRIG, which would be required of future members (SHIRBRIG 2001c). First, SHIRBRIG members had to be small or middle powers. In an October 2003 interview, Mette Kjuel Nielsen, the former Chair of the Steering Committee, claimed that "SHIRBRIG is the perfect vehicle for smaller and middle-sized, like-minded nations. Deploying with countries you know in advance enhances the security."[14] The Steering Committee did not expect the permanent members of the UN Security Council to express any interest in joining SHIRBRIG, since they were capable of deploying brigade-size units on their own. But the Steering Committee did view the political and military support of the great powers as useful for the successful deployment of SHIRBRIG. For example, Nielsen argued that SHIRBRIG would have benefited if a strategic transportation agreement with the United States could have been arranged.[15]

Second, members of SHIRBRIG had to also participate in UNSAS and have peacekeeping experience. Third, SHIRBRIG members had to be able to pay for their participation in the brigade. The costs of participation were minimal. Members paid for the training and preparation for deployment of their own national units, but once deployment began, the UN paid for all of SHIRBRIG's expenses. The SHIRBRIG members shared the costs of the PLANELM and the Steering Committee. For example, the budget of the PLANELM was projected to be $440,000 in 2002, and the costs were shared by the ten full members (SHIRBRIG 2001c). Fourth, members needed to make capable units available to SHIRBRIG at the required level of readiness. Fifth, the Steering Committee encouraged diversity in SHIRBRIG's membership; hence, new members had to be representative of different regions of the world. Finally, prospective members had to be willing to accept each of the four documents which constituted the framework of SHIRBRIG.

SHIRBRIG in action

During the ninth meeting of the Steering Committee in Stockholm on October 7–8, 1999, the member states decided that SHIRBRIG had reached the necessary level of readiness for UN deployment (SHIRBRIG 2001c). On December 17, 1999, the Steering Committee informed UN Secretary-General Kofi Annan that SHIRBRIG would be made available to the United Nations as of the end of January 2000. The UN made immediate use of SHIRBRIG's services. On April 26, 2000, the UN inquired informally if SHIRBRIG would be available for a possible deployment as part of the United Nations Interim Force in Lebanon (UNIFIL), which had been stationed in Southern Lebanon since March 1978. SHIRBRIG conducted a fact-finding mission in the UNIFIL area of operations on May 16–17, 2000, and then issued a declaration on June 13th that it was available for deployment in Southern Lebanon. As Hans Hækkerup recalled:

> DPKO asked if SHIRBRIG could go to South Lebanon. After hard work I managed to put together a battalion ([over] 1,000 soldiers) from Poland, Denmark and Lithuania. Canada and the Netherlands did not want to participate, it was too risky. The whole situation changed because the Israelis left South Lebanon overnight, and DPKO decided they didn't need more troops.[16]

Hækkerup also noted that some members of the DPKO may have had devious motives in requesting SHIRBRIG's deployment to Lebanon:

> In DPKO there certainly also were people who wanted to "call the bluff" by sending SHIRBRIG to a mission it was not ready for. With Southern Lebanon they nearly succeeded.[17]

But on June 16th, SHIRBRIG received another informal inquiry from the UN if the brigade would be available for a mission in Ethiopia and Eritrea. The two countries, which had engaged in warfare in 1998–9 due to a border dispute, had resumed fighting on May 12, 2000 (UN Department of Public

Information 2005 [hereafter UNDPI]). The Organization of African Unity (OAU), with the assistance of the European Union and the United States, had brokered the Agreement on Cessation of Hostilities between Ethiopia and Eritrea, which was signed in Algiers on June 18th by both parties to the conflict.

In order to preserve the peace, the parties called on the UN and the OAU to establish a peacekeeping operation. On June 30th, UN Secretary-General Annan informed the Security Council that he was going to send liaison officers to Addis Ababa and Asmara, to be followed by the gradual deployment of one hundred military observers to each country over the next two months, until a peacekeeping operation could be assembled (UNSG 2000b). The Security Council then issued Resolution 1312 (UN Security Council 2000a [hereafter UNSC]) on July 31st, announcing the creation of the UN Mission in Ethiopia and Eritrea (UNMEE). UNMEE was authorized to deploy one hundred military observers and a civilian support staff until a peacekeeping force could be established. Its mandate was to perform a liaison function with the parties to the conflict, visit the parties' military headquarters and other units in the mission's area of operations, implement the mechanism for verifying the cessation of hostilities, prepare for the creation of a Military Coordination Commission as stipulated in the Agreement on Cessation of Hostilities, and participate in planning for a future peacekeeping mission.

The Security Council asked Secretary-General Annan to proceed with preparations for a peacekeeping mission. In his report to the Security Council on August 9, 2000, the Secretary-General described the mandate of an expanded UNMEE and recommended that 4,200 military personnel, including 220 military observers, three infantry battalions, and support units, be deployed to monitor the ceasefire and the border delineation between Ethiopia and Eritrea (UNSG 2000c). On September 15th, in response to the Secretary-General's report, the Security Council issued Resolution 1320 authorizing UNMEE to deploy up to 4,300 troops until March 15, 2001 (UNSC 2000b).

Upon receiving authorization from the Security Council, SHIRBRIG sprung into action (SHIRBRIG 2001c, 2007a, 2010a). The SHIRBRIG Commander's conference, held in Norway on September 25–29, 2000, focused on the Horn of Africa. On October 10th, the PLANELM provided the Commander with a mission analysis briefing. Members of the PLANELM left Denmark for Asmara, Eritrea on November 16th, in order to assist the UN in forming the Force Headquarter. The deployment of SHIRBRIG units to Ethiopia and Eritrea continued through November and December. The total SHIRBRIG deployment consisted of a headquarters unit, based in Asmara, which featured the active participation of each SHIRBRIG member; an infantry battalion made up of Dutch and Canadian soldiers; and a headquarters company from Denmark. The SHIRBRIG Commander, Brigadier-General P. C. Cammaert from the Netherlands, was appointed as UNMEE Force Commander.

In the meantime, the parties in conflict were engaged in peace talks brokered by President Abdelaziz Bouteflika of Algeria. On December 12, 2000, Ethiopia and Eritrea signed a comprehensive Peace Agreement in which they promised to terminate military hostilities permanently and refrain from threatening or using force against each other (UNDPI 2005). In May and June of 2001, SHIRBRIG pulled out of Ethiopia and Eritrea, having completed its peacekeeping tasks successfully during the six-month deployment.

Following its withdrawal from Ethiopia and Eritrea, SHIRBRIG entered a reconstitution phase which lasted around seven months. In this phase, the Steering Committee of SHIRBRIG evaluated the lessons that were learned from the UNMEE mission (SHIRBRIG 2001c). First, SHIRBRIG had to increase the size of its brigade pool in order to provide for greater redundancy and geographic representation. Second, SHIRBRIG members needed to ensure that their units were available for potential deployments. Hans Hækkerup commented that "it is crucial that SHIRBRIG is deployed again and that more countries participate with contingents and not only with staff officers."[18]

Third, SHIRBRIG had to improve its direct liaison and early integration with the United Nations. Fourth, SHIRBRIG had to adjust its force structure in order to improve its headquarters unit in the mission area. Finally, SHIRBRIG concepts and key documents needed to be updated to incorporate the lessons learned from the UNMEE experience. Peter Viggo Jakobsen suggested that SHIRBRIG "should be expanded to include operations beyond traditional peacekeeping. The full spectrum of peace operations should be included and [SHIRBRIG] should have a strategic lift capacity."[19]

On October 15, 2001, the Presidency of the SHIRBRIG Steering Committee informed the Security Council that SHIRBRIG would once again be made available to the UN as of January 1, 2002 (SHIRBRIG 2001c). Although SHIRBRIG never deployed again on the same scale as its participation in UNMEE, the brigade remained active in the years that followed (SHIRBRIG 2010). In March 2003, SHIRBRIG provided the Economic Community of West African States (ECOWAS) with a planning team, in order to assist with the preparation of a peacekeeping mission for Côte d'Ivoire, following an armed revolt by disgruntled soldiers that began the previous September. In addition, twenty SHIRBRIG personnel were sent to Liberia in September 2003, to serve as the nucleus of the interim headquarters of the UN Mission in Liberia (UNMIL), which was mandated to enforce the Comprehensive Peace Agreement that ended the bloody civil war in August. The SHIRBRIG team was redeployed to Denmark in November 2003, one month after UNMIL assumed responsibility for peacekeeping duties from the previous ECOWAS mission in Liberia (SHIRBRIG 2007b).

From July 2004 until February 2005, fourteen members of the brigade's PLANELM were deployed as part of the UN Advance Mission in Sudan (UNAMIS), a special political mission initially mandated to deal with the

North-South conflict that had ravaged the country for more than two decades, and later assigned an additional responsibility for the crisis-stricken region of Darfur. SHIRBRIG also participated in the subsequent UN Mission in Sudan (UNMIS) from April to November 2005 (SHIRBRIG 2006). The brigade provided UNMIS with the core of the Force Headquarters in Khartoum, the Joint Military Coordination Office in Juba, and the Integrated Support Services. Moreover, SHIRBRIG deployed a headquarters and a security unit from Italy in Sudan, and the Commander of SHIRBRIG, the Canadian Brigadier-General Gregory Mitchell, served as the Deputy Force Commander of UNMIS, while Swedish Colonel A. S. Lund was appointed UNMIS Chief of Staff.

SHIRBRIG activities in Sudan continued in 2006, when the DPKO requested assistance from the brigade in planning the establishment of a divisional headquarters in Darfur, where UNMIS peacekeepers were preparing to replace the African Union Mission in the Sudan (AMIS). In 2007, SHIRBRIG was requested to help the African Union (AU) in planning for the African Union Mission in Somalia (AMISOM). The following year, SHIRBRIG was asked to assist the UN and the EU with the UN Mission in the Central African Republic and Chad (MINURCAT) (SHIRBRIG 2010a).

It is surprising that SHIRBRIG was only used sparingly and in a limited capacity following its successful participation in UNMEE. According to Peter Viggo Jakobsen, the main reason for the infrequency of SHIRBRIG deployments may have been the fact that "it is very difficult to find suitable 'traditional peacekeeping [operations]'."[20] Mette Kjuel Nielsen concurred that there was a lack of appropriate missions for SHIRBRIG participation, but also added that the UN Security Council showed little interest in engaging in traditional peacekeeping operations in recent years, and that the key SHIRBRIG members—Canada, Denmark, and the Netherlands—were deeply involved in other missions.[21]

Sadly, international neglect sealed the fate of SHIRBRIG. Despite the incredible efforts of the middle powers to establish and deploy SHIRBRIG, the brigade terminated all participation in UN operations by December 31, 2008 (SHIRBRIG 2010a). The following year, the PLANELM ended its operations with the AU. On June 30, 2009, SHIRBRIG passed into history.

The US reaction to the SHIRBRIG initiative

The United States acquiesced to the formation and deployment of SHIRBRIG. Since the end of the Cold War, the United States has encouraged the efforts of the United Nations to reform its peacekeeping operations. Even former President Ronald Reagan, who was an ardent opponent of the UN during his presidency, supported the reform of UN peacekeeping. In a December 1992 speech, Reagan called for the creation of a standing UN "army of conscience," that would be capable of using force to establish humanitarian sanctuaries (Albright 1993).

The Democrats have also perceived the need for reforms in UN peacekeeping missions. In August 1992, US presidential candidate Bill Clinton expressed his support for the establishment of a voluntary UN rapid deployment brigade (Langille 2000). After the Clinton administration came into power, US Secretary of State Warren Christopher notified UN Secretary-General Boutros Boutros-Ghali in February 1993 that the United States would back the development of a UN rapid response force. It should be noted that the Clinton administration did not support the formation of a standing UN army. This policy position echoed the sentiments of the US Congress and past presidential administrations. Instead, the Clinton administration proposed that the UN should set up a rapidly deployable headquarters team, a logistics support unit, a database of national military units that would be available for deployment, a trained civilian reserve corps, and a modest airlift capability. The Clinton administration believed that by establishing a capability for rapid reaction, the UN would be able to prevent the massive violations of human security that could result from delays in deployment following the authorization of a mission (Taylor, Daws, and Adamczick-Gerteis 1997).

The United States is a member of UNSAS and has backed the recommendations of the "Brahimi Report" on UN peacekeeping reform (US Department of State 2000). The Brahimi Report was issued on August 23, 2000, by the independent Panel on United Nations Peace Operations, which was chaired by Ambassador Lakhdar Brahimi of Algeria, and consisted of ten peacekeeping experts appointed by UN Secretary-General Kofi Annan. Among the report's many recommendations was a call for UN peacekeeping missions to become rapidly deployable (Panel on United Nations Peace Operations 2000). The United States also participated in UNMEE, albeit in a token manner by contributing a total of seven military observers (US Department of State 2001). The US government did not issue any official statements regarding the establishment and deployment of SHIRBRIG. But one can conclude that Washington acquiesced to the SHIRBRIG initiative, because the brigade corresponded to American preferences by enhancing the rapid response capability of the UN without being a standing army.

Conclusion

The SHIRBRIG initiative demonstrates how the middle powers were able to exercise strong leadership on an issue of human security. The middle powers addressed the need for a UN rapid response capability, the lack of which had dire consequences for the security of people in numerous conflicts around the globe. This human security initiative illustrates the effectiveness of middlepowermanship. Canada, Denmark, and the Netherlands used their technical expertise in the area of peacekeeping to launch an initiative to create SHIRBRIG, and employed their entrepreneurial skills to build a coalition—the Friends of Rapid Deployment—that would support and promote their proposal. Although SHIRBRIG's existence was brief, it was deployed successfully on a large scale as part of UNMEE, as well as in a lesser capacity elsewhere in Africa. Mette Kjuel Nielsen emphasized the essential contributions of middle power leadership for the creation of a UN rapid response capability:

> In my opinion, Denmark and Canada have been the core [leaders], together with Holland. Without the driving force and total commitment of the Danish Defense Minister Hans Hækkerup (who fostered the idea), SHIRBRIG would not have been a reality. He convinced the UN Secretary General Kofi Annan of the idea, he pushed for the letters of intent to be signed, and when it came to all out difficulties over the first deployment, he spoke on the phone and held a number of meetings with his colleagues to make it happen.[22]

The SHIRBRIG initiative did not threaten the core national interest of the United States. Because the initiative created a standby brigade rather than a standing army, the United States acquiesced to the formation of SHIRBRIG. The disbanding of the brigade in 2009 does not indicate any failure on the part of this middle power initiative. Rather, the changing nature of conflict was to blame, as the demand for peace enforcement in the contemporary era outweighs the need for traditional peacekeeping. While the middle powers have the resources and skills to carry out traditional peacekeeping, revamping SHIRBRIG to make it capable of contributing to peace enforcement would have probably required the participation of at least one Western great power, whether the United States, France, or the United Kingdom. The risk of surrendering control of SHIRBRIG to one of these states was a price that the middle powers found to be too high to pay.

Notes

1 David Cox expressed a different view that "peace-enforcement units [are] a mid-point between traditional UN peacekeeping and Chapter VII-style enforcement actions" (Cox 1993, 10).

2 The combined activities of the UN Temporary Executive Authority (UNTEA) and the UN Security Force (UNSF) in assisting the transfer of West New Guinea (West Irian) from Dutch to Indonesian administration, from October 1962 until April 1963, resembled a peace-building mission, although UNTEA and UNSF existed as two separate operations in West Irian. See Asada 1995.

3 This figure contrasted sharply with Brian Urquhart's estimate that a light infantry, rapid response brigade consisting of five thousand troops would cost around $380 million per year to maintain and equip. See Urquhart et al. 1997.

4 In his 1998 article with François Heisbourg, Brian Urquhart recommended that the rapid response brigade should consist of ten thousand soldiers rather than the five thousand he originally suggested. In contrast, Robert Johansen adviced that a rapidly deployable police or constabulary force consisting of ten to twenty thousand volunteers should be established, and allowed to grow "to ten times that size if demands for UN peacekeeping continue to rise" (Johansen 1998, 106).

5 E-mail interview with Dr. Peter Viggo Jakobsen, the Head of the Department of Conflict and Security Studies at the Danish Institute for International Studies, Copenhagen, Denmark, September 29, 2003.

6 E-mail interview with Mr. Hans Hækkerup, the former Danish Minister of Defense (1993–2000), Beijing, China, September 2, 2004.

7 The middle power participants were Argentina, Belgium, Canada, the Czech Republic, Denmark, Finland, Ireland, the Netherlands, New Zealand, Norway, Poland, and Sweden. See Langille 2000, fn. 24.

8 Ministers from Australia, Canada, Denmark, Jamaica, the Netherlands, New Zealand, Nicaragua, Senegal, and Ukraine participated in the meeting. See Langille 2000, fn. 33.

9 The members of FORD were Argentina, Australia, Bangladesh, Brazil, Canada, Chile, Denmark, Egypt, Finland, Germany, Indonesia, Ireland, Jamaica, Japan, Jordan, Malaysia, the Netherlands, New Zealand, Nicaragua, Norway, Poland, Senegal, South Korea, Sweden, Ukraine, and Zambia. See Langille 2000.

10 E-mail interview with Dr. Peter Viggo Jakobsen, September 29, 2003.

11 E-mail interview with Mr. Hans Hækkerup, September 2, 2004.

12 Although Argentina signed all four SHIRBRIG documents and became a full member of the brigade, it would later suspend its membership in SHIRBRIG.

13 Hungary was an Observer Nation as well, but suspended its observer status in SHIRBRIG in November 2007.

14 E-mail interview with Ms. Mette Kjuel Nielsen, the Head of the Department for Russia, CIS, OSCE and the Balkans at the Danish Ministry of Foreign Affairs, and the former Danish Deputy Permanent Secretary of Defense (1998–2001), Copenhagen, Denmark, October 11, 2003.

15 Ibid.

16 E-mail interview with Mr. Hans Hækkerup, September 2, 2004.

17 Ibid.

18 Ibid.

19 E-mail interview with Dr. Peter Viggo Jakobsen, September 29, 2003.

20 Ibid.

21 E-mail interview with Ms. Mette Kjuel Nielsen, October 11, 2003.

22 Ibid.

CHAPTER FOUR

Banning antipersonnel landmines: the Ottawa Process

The success of the "Ottawa Process" in achieving a swift ban on the use, stockpiling, production, and transfer of antipersonnel landmines (APLs) was astonishing. Starting in October 1996, with Canadian Foreign Minister Lloyd Axworthy's call for an international convention banning APLs, it took less than two-and-a-half years for the Ottawa Convention to come into force in March 1999. The results of the Ottawa Convention were also amazing. By June 2000, the number of states that produced APLs had dropped by two-thirds, and thirty-three out of thirty-four APL exporting countries had issued either a ban or a moratorium on exports (Gwozdecky and Sinclair 2001). Moreover, APLs had become discredited as weapons of war.

The worldwide ban on antipersonnel landmines would have never been achieved without the skilled leadership of the middle powers. As this chapter shall illustrate, the APL initiative is a textbook case of fast-track diplomacy in action. The first section of the chapter discusses the international trade in antipersonnel landmines, the usefulness of APLs for national militaries, and the horrible consequences the global proliferation of APLs has had for civilians. This is followed by an analysis of international action on the issue of APLs, from nineteenth-century attempts at regulating landmines through the Ottawa Process. The reaction of the United States to the Ottawa Process and the astounding results of the APL initiative are also explored.

The proliferation of antipersonnel landmines

The international trade in landmines

In a 1993 report, the Arms Project of Human Rights Watch estimated that five to ten million APLs had been manufactured in recent decades (Arms Project of Human Rights Watch and Physicians for Human Rights 1993 [hereafter HRW/PHR]). The global production of APLs, excluding delivery systems and accessories, was assessed by the Arms Project to be worth \$50–\$200 million annually. Evidence suggested that China, Italy, and the former Soviet Union were the largest producers of APLs, based on the numbers of their mines found around the world. If antipersonnel submunitions were included, the United States would have probably ranked as the largest or second-largest producer.

More than 340 types of landmines were developed by over one hundred companies and government agencies in fifty-two countries (Vines 1998). The 1993 Arms Project report ranked the leading APL developers over the prior quarter-century in terms of the number of APL models they had created. With thirty-seven APL models, the United States was the leading innovator in the landmine industry, followed closely by Italy (thirty-six models) and the former Soviet Union (thirty-one). Other major players in the landmine industry were Sweden (twenty-one models), Vietnam (eighteen), East and West Germany combined (eighteen), Austria (sixteen), the former Yugoslavia (fifteen), France (fourteen), China (twelve), and the United Kingdom (nine).

The international landmine trade tended to be a confidential affair, with few states releasing import, export, and procurement data. Since direct sources of data were limited, the Arms Project drew on landmine deployment data in order to discern which states were producing and exporting APLs. At least forty-one companies and government agencies in twenty-nine countries had exported APLs (HRW/PHR 1993). Most experts concurred that China, Italy, and the former Soviet republics were leading exporters, although there were disagreements about each country's individual ranking. A 1992 report by the US Defense Intelligence Agency and the US Army Foreign Science and Technology Center (hereafter USDIA/USAFSTC) claimed that the former East Germany, Italy, the former Soviet Union, and the former Czechoslovakia were the sources of the majority of landmines purchased by developing countries from the 1970s until the 1990s. The report also highlighted China, Egypt, Pakistan, and South Africa as new players on the APL export market (USDIA/USAFSTC 1992). According to the Arms Project, Belgium, Chile, Greece, Israel, Portugal, Singapore, Spain, and the former Yugoslavia had also become significant APL exporters by the early 1990s (HRW/PHR 1993).

The US State Department downplayed the importance of American APL exports, arguing both that the United States was a selective exporter of limited quantities of landmines, and that less than fifteen percent of the APLs in countries plagued with uncleared landmines originated in the United States (US Department of State 1993). But the Arms Project claimed that a figure of fifteen percent would still rank the United States among the top five exporters of landmines worldwide (HRW/PHR 1993). Since 1969, the United States has exported over 4.3 million APLs, with the most exports, 1.4 million APLs, occurring in 1975 (Vines 1998). Exports of APLs to Cambodia, Chile, and Iran in 1975 made up one-third of all reported APL sales to foreign militaries in the twenty-four year period from 1969 to 1993. Other major purchasers of US landmines were El Salvador, Malaysia, Saudi Arabia, and Thailand. American sales to foreign militaries were considerably less in the 1980s, numbering around seventy thousand APLs. Half of these mines were sold to El Salvador, while Lebanon and Thailand were other leading customers.

Alex Vines (1998) cautioned that the U.S. government data provide no information about the licensed production of American APL models abroad, the unauthorized copying of American APL models in other states, the covert shipments of APLs to insurgent groups during the Cold War, and the deployment of APLs by the US military in conflicts such as the Vietnam War and the Persian Gulf War. The Arms Project recognized that the United States has not been a significant APL exporter since the 1970s (HRW/PHR 1993). Furthermore, there are alternative explanations as to why contemporary mine clearers appear to encounter so many American-made APLs, including the possibility that many of those mines have been deployed since the 1970s, or that some of the mines are copies of American models that are produced by other states. In addition, countries which originally purchased American APLs may have resold them to other states, and numerous APLs were also shipped covertly by the US government to rebel groups in countries like Afghanistan, Angola, Cambodia, and Nicaragua. Nevertheless, the American landmine industry was a profitable business. Between 1985 and 1995, Alliant Techsystems, the largest manufacturer of APLs in the United States made $350 million in landmine sales, while its subsidiary Accudyne, the third largest producer, raked in $150 million (Capellaro and Cusac 1997).[1]

The military use of landmines

Military forces hail the efficacy of landmines as defensive weapons. Minefields provide a semipermanent barrier that can be used to protect national borders, military and economic assets, and soldiers (Roberts and Williams 1995). In an era when conventional warfare requires speedy maneuvers, landmines can hinder the movement of an enemy and deny it access to key tactical positions (HRW/PHR 1993). Landmines may also be used to direct enemy soldiers to move into a vulnerable area, where they can be defeated more easily in battle. In order to breach a minefield, soldiers must engage in the perilous and time-consuming task of mine clearance, thus, the deployment of APLs and antitank mines can slow down the advance of an enemy army. Some landmines, such as the Swedish L1-11 and the US M14, were designed to maim victims without killing them, so that high casualty rates would burden an enemy's medical facilities as well as reduce the morale of the troops. Landmines have been referred to as a "force-multiplier," because their effects enhance the usefulness of other weapons (Roberts and Williams 1995). Proponents of landmines tend to emphasize their utility and cost-effectiveness as a weapon. They argue that if landmines are directed toward military targets, civilian casualties can be limited.

But military forces were increasingly using landmines as offensive weapons. In order to disrupt an enemy's logistics, scatterable mines could be dropped by aircraft between an advancing army and its supply base. Landmines could also be used by an attacking army in order to force a defending army to fight from a tactically inferior position. An advancing army could use landmines to secure its flanks, seal off approach routes, strengthen a temporary defense, or halt a counter-attack (HRW/PHR 1993). In contrast to the defensive use of landmines as tactical weapons in support of other weapon systems, the offensive deployment of landmines could turn them into strategic weapons that can overcome the low force-to-space ratio that characterizes guerrilla warfare. Shawn Roberts and Jody Williams summarized the dreadful purposes for which landmines have been utilized as offensive weapons:

> Just as landmines have been used to deny access of terrain to enemy troops, they have been deployed to depopulate whole sections of countries, to disrupt agriculture, to interrupt transportation, to damage economic infrastructure, and to kill and maim thousands of innocent civilians. Landmines have been used by both regular and irregular armies to undermine the social and economic fabric of society. They have been deployed to make vital economic assets useless and cripple the economic and social redevelopment of these countries after the wars are over. (Roberts and Williams 1995, 5)

A global contamination

In 1997, it was estimated that more than one hundred million landmines were deployed worldwide (Morrison and Tsipis 1997). Moreover, between two and five million new landmines were being laid each year (American Medical Association 1997 [hereafter AMA]). Most countries have been affected by the scourge of APLs, whether they have experienced landmine incidents on their own soil, or have had peacekeepers and civilian aid workers killed by landmines abroad. The most adversely affected region has been Southern Africa. According to a 1998 study by Alex Vines, there were at least twenty million mines in the region, and eleven of the fourteen member states of the Southern Africa Development Community (SADC) had reported landmine incidents (Vines 1998). Since the first mine casualty was recorded in Angola in 1961, there have been over 250,000 victims of landmines in Southern Africa.

The 1993 report by the Arms Project provided a regional overview of the APL contamination (HRW/PHR 1993). The whole region of Sub-Saharan Africa contained as many as thirty million mines laid in eighteen countries. The most severely mined countries were Angola, Djibouti, Eritrea, Ethiopia, Malawi, Mozambique, Somalia, and Sudan.[2] Eight Middle Eastern countries had an estimated total of seventeen to twenty-four million landmines. Kuwait, Iran, and Iraq were the most heavily mined countries. There were between fifteen and twenty-three million landmines in eight states in East Asia, of which Cambodia and Vietnam were the most severely affected.[3] South Asia had between thirteen and twenty-five million landmines. The vast majority of these mines were in Afghanistan, and along the borders between Afghanistan, China, India, and Pakistan. Although Europe had relatively less of a landmine crisis, with three to seven million mines in thirteen countries, it was the region that had experienced the most rapid increase in the number of mines. This was due mainly to the conflicts in the Balkans, as the landmine problem was most severe in the former Yugoslav states of Bosnia-Herzegovina and Croatia. There were also between three hundred thousand and one million landmines in eight countries in Latin America. The worst trouble spots were in El Salvador and Nicaragua (HRW/PHR 1993).

The devastating effects of antipersonnel landmines

Antipersonnel landmines are also known as "weapons of indiscriminate mass destruction," due to the deadly consequences they have for both combatants and noncombatants (Capellaro and Cusac 1997). In 1997, up to two thousand people around the world were being maimed or killed by landmines every month (Garner 1997). This rate of landmine-related casualties had doubled since 1980 (AMA 1997). According to the International Committee of the Red Cross (ICRC), landmines had claimed more victims than either chemical or nuclear weapons (Garner 1997).

Landmines leave a significant psychological impact on their victims, and place tremendous demands on a state's health care system. Amputation or blindness resulting from an APL may very well end the working life of a peasant. The victim's children are often forced to leave school so that they can work full-time in order to supplement the family income. Most medical facilities in mine-contaminated countries are poorly equipped to deal with landmine casualties, which results in a higher rate of amputations and deaths (Roberts and Williams 1995). Prostheses are prohibitively expensive for many victims. The ICRC recommended that a prosthesis should be replaced every six months for a child, and every three to five years for an adult. UN Secretary-General Boutros Boutros-Ghali (1994) calculated that a ten-year-old victim with a life expectancy between fifty to sixty years would need around twenty-five prostheses in their lifetime. With each artificial limb costing approximately $125, the victim would need to spend around $3,125 in their lifetime on prostheses.

In addition to the tragic losses in human lives, landmines have economic and social costs. The presence of APLs hinders a country's efforts to rebuild following a war, since mine clearance is a dangerous and time-consuming process. Roads, bridges, power lines, schoolyards, and farmlands are popular targets of mine-layers (Fields 2001). The most severe proliferation of landmines is in less-developed countries, which also tend to be the most dependent on the use of land for the purpose of economic development (Faulkner 1997). The random and unmapped placement of APLs renders prime agricultural land unusable and uninhabitable. The grazing of livestock becomes hazardous because herds, and the people who tend them, may wander onto unmarked minefields in search of better feeding grounds. The collection of drinking water and firewood is especially perilous when water sources and forests are mined. Women and children are the ones who often perform these tasks, and thus become victims of APLs.

Countries with a limited infrastructure may be the most affected by the deployment of APLs (Roberts and Williams 1995). The mining of dams and electrical installations may hamper the ability of a country to generate enough power for reconstruction after a conflict. Markets are disrupted because people find it too dangerous to transport goods and services on roads that have been mined. These disruptions have a negative impact on employment, and produce inflation due to scarcities in goods and services. Furthermore, the presence of landmines amplifies the effects of droughts, because humanitarian relief agencies find it too treacherous to deliver food aid over mine-infested roads. In fact, countries with acute landmine problems tend to become an economic burden on the international community; of the sixteen countries who received UN humanitarian assistance in 1993, thirteen were contaminated with landmines (Roberts and Williams 1995). With the exception of Kuwait, states with severe landmine problems have relied on financial assistance from the global community to fund demining programs.

The heavy deployment of landmines has harsh effects on the environment (Roberts and Williams 1995). Wherever the placement of landmines has reduced the amount of available agricultural land, populations are forced to overutilize the remaining land, thereby hastening its degradation. Populations may also turn from mined agricultural lands to the forests for their livelihood, thus causing an accelerated rate of deforestation that impacts the ecological balance of flora and fauna. Moreover, landmines cause the displacement of populations, as people flee heavily mined areas. These refugees often head to overcrowded cities, where they live in miserable conditions without finding employment. The resettlement of refugees after the end of a conflict may produce many APL casualties, as people return to their villages and lands unaware that these locations have been mined in their absence. Landmines also kill many wild animals, some of which are rare species. In addition, the mining of agricultural lands may force people to turn to hunting in order to ensure their food supply, further endangering the survival of animal species (National Wildlife Federation 2000).

Postconflict demining activities are extremely dangerous. In Kuwait, where around seven million mines were laid during the Gulf War, more than eighty mine clearers have been killed or injured. More than thirty deminers have lost their lives in Afghanistan. Humanitarian mine clearance requires the removal of every mine in a minefield. In order to be deemed successful, the mine clearance rate must be over ninety-nine percent, and preferably over 99.9 percent (Boutros-Ghali 1994). There has been little improvement in mine clearance techniques since the 1940s, as more money has been spent instead on devising new means for militaries to breach minefields. As a result, armed forces possess the wrong equipment and lack the necessary training for humanitarian demining. The removal of APLs is done by hand, with one specialist using a metal detector, and another specialist down on their knees probing the ground with a stick. Instead of achieving a ninety-nine percent clearance rate, the detection equipment that is currently used is only sixty to ninety percent effective in locating APLs that are made with a minimum of metal. Mines that are manufactured from plastic cannot be detected with this equipment. Furthermore, mine clearance is often hampered by booby traps that were set for the purpose of preventing demining. Locating APLs is even more difficult when minefields have not been mapped, as required under the 1980 Convention on Certain Conventional Weapons (CCW). Even if maps have been made, the exchange of territory during battle may result in the unrecorded placement of new mines on the minefields by enemy soldiers. In addition, changes in weather conditions, such as floods, may move mines around (Roberts and Williams 1995).

The primary responsibility for demining lies with the country affected (Boutros-Ghali 1994). The UN encourages the training of local civilians as deminers. Funding for many of the mine clearance programs around the world is provided by the UN, under two conditions: national governments must grant their consent, and security arrangements must be made for UN personnel who remove APLs in militarily sensitive areas. Demining is a very expensive activity because it is labor-intensive. For example, a UN-managed mine clearance program in Afghanistan, which deploys around two thousand deminers, costs approximately twelve million dollars annually. It is estimated that the cost of any mine clearance program, including training, support, and logistics, ranges from three hundred to one thousand dollars per mine (Boutros-Ghali 1994). This is an exorbitant amount when compared with the actual cost of an APL. Most APLs are priced less than twenty-five dollars, while some are even cheaper than three dollars.

The campaign to ban antipersonnel landmines

Earlier legal restrictions on landmines

The legal foundations for the international regulation of APLs date to the nineteenth century. In 1868, seventeen countries signed the Declaration of St. Petersburg, which established three principles (Baxter 1977). The first principle made a distinction between combatants and noncombatants during war, and stipulated that the latter may not become the targets of attack. The second principle stated that force should only be applied to the point that enemy troops are disabled, and that weapons should not be used to aggravate the suffering of the wounded or render their death inevitable. Finally, the third principle prohibited the antipersonnel use by the military of any explosive or inflammable projectile of less than four hundred grams. Although only the states parties were legally bound by the Declaration of St. Petersburg, it is widely viewed as international law and therefore obligatory for all states.

The 1899 Hague Peace Conference adopted the Convention with respect to the Laws and Customs of Warfare on Land. The convention established the principle that belligerents have a limited right when choosing weapons to injure the enemy, and proscribed the use of poison and arms, projectiles, or material that may cause superfluous injury. The convention also prohibited the treacherous killing of individuals belonging to the hostile country or military. While treacherous killing was initially viewed as the deceitful use of a flag of truce or the feigning of disablement in order to kill, some have claimed that it includes the deployment of mines and booby traps (Baxter 1977). At the Second Hague Peace Conference of 1907, the inaccurate English translation of the original French text changed the reference from weapons causing "superfluous injury" to weapons producing "unnecessary suffering." Nevertheless, the objective of the Hague Peace Conferences was "to confirm the standard of St. Petersburg and to reaffirm the principle that in the employment of weapons humanitarian considerations and military necessity shall be balanced" (Bring 1987, 277).

Following the horrors of the Vietnam War, the ICRC and a few NGOs began to pressure governments to examine the issue of weapons that are indiscriminate or cause unnecessary suffering, such as landmines (Williams and Goose 1998). The 1973–77 Geneva Diplomatic Conference on Humanitarian Law produced two Additional Protocols to the 1949 Geneva Conventions, of which Protocol I reiterated the prohibition on weapons and methods of warfare that may cause superfluous injury or unnecessary suffering.[4] But Ove Bring (1987), a legal adviser for the Swedish Ministry of Foreign Affairs, argued that general proscriptions, such as those expressed in Protocol I, are not very useful, because states will often consider a particular weapon as not prohibited unless it is named explicitly. The United Nations

Conference on Prohibitions or Restrictions on the Use of Certain Conventional Weapons Which May be Deemed to be Excessively Injurious or to Have Indiscriminate Effects was held in September 1979 and September 1980 (Bring 1987; Maresca and Maslen 2000; Szasz 1980). On October 10, 1980, the conference adopted the Convention on Certain Conventional Weapons.[5] The CCW initially contained three annexed protocols, each of which is concerned with a particular type of weapon. Protocol I bans the use of weapons which injure by fragments that are not detectable by X-rays. Protocol II, also known as the Mines Protocol, restricts the use of mines and booby-traps. Protocol III limits the use of napalm and other incendiary weapons. In 1995, Protocol IV restricting blinding laser weapons was added to the CCW, and Protocol V dealing with the problem of explosive remnants of war was adopted in 2003.

Both antipersonnel and antitank mines are regulated under Protocol II (Maresca and Maslen 2000). The indiscriminate use of mines is proscribed, as is the manual placement of mines in populated areas unless there is active or imminent combat, or the minefields are clearly marked. Protocol II bans the deployment of mines that are remotely deliverable, with the exception of situations where these mines are used in the vicinity of a military target, and either the locations of the mines are recorded accurately or the mines possess a neutralizing mechanism. Moreover, Protocol II requires the mapping of minefields. By November 2011, there were 114 states parties to the CCW, and ninety-three states parties to Protocol II (UN Office at Geneva 2011).

Ove Bring criticized Protocol II as being "insufficient in the sense that it does not effectively deal with the question of 'material remnants of war'" (Bring 1987, 278). Article 9 of the Protocol stipulates that after hostilities have ceased, states shall cooperate with each other and with international organizations, in an attempt to derive an agreement on assistance to remove or disable mines and booby traps that were laid during the conflict. But this does not guarantee that postwar mine clearance will remove all the APLs that were deployed during the conflict. Michael Matheson argued that the Mines Protocol has considerable shortcomings, and that it is "a Western proposal and basically codified the practices already being observed by US and other Western military forces in the use of these weapons" (Matheson 1997, 159). Furthermore, despite the inclusion of Protocol II in the Convention on Certain Conventional Weapons, landmines received relatively little attention during the negotiation of the CCW (Maresca and Maslen 2000). The international community was more troubled at the time by the problem of incendiary weapons; hence, Sweden and other middle powers devoted their energies toward achieving prohibitions on the use of incendiaries (Baxter 1977).

But by the end of the Cold War, the global community switched its focus to the issue of landmines, as it had become clear that the conflicts of the 1980s had produced a great number of civilian casualties from the indiscriminate use of APLs (Matheson 1997). The ICRC and the NGOs that worked in mine-infested countries were the first to draw attention to the humanitarian crisis. In response to the increasing number of casualties from landmines, the ICRC held a symposium on antipersonnel mines in Montreux, Switzerland on April 21–23, 1993. The purpose of the meeting was to assess the scope of the APL problem, evaluate possible courses of action to reduce APL use, and review the means of caring for mine victims. Participation in the symposium was broad, and included APL specialists, manufacturers, military strategists, doctors, rehabilitation specialists, legal advisers, deminers, and representatives of NGOs. Copies of the report that was produced by the symposium were sent to all governments in August 1993. According to Louis Maresca and Stuart Maslen, "the report on the Montreux Symposium became an important source of reference for the ICRC, nongovernmental organizations and governments in their future activities in pursuit of a ban treaty" (Maresca and Maslen 2000, 129).

In February 1993, the government of France requested that UN Secretary-General Boutros Boutros-Ghali convene a conference of the parties to the CCW, under the auspices of the Conference on Disarmament (CD), in order to review the provisions of the convention (Matheson 1997). The parties endorsed this request in December 1993 and called on the Secretary-General to establish a group of governmental experts to prepare for the conference. The group of governmental specialists was set up and four sessions were held in Geneva between February 1994 and January 1995. The recommendations of the group for a revision of the CCW were then discussed in a Review Conference that spanned three sessions. The first session was held in Vienna in September and October 1995, the second session in Geneva in January 1996, and the third session once again in Geneva in April and May 1996. Although there was a widespread agreement among states that the Mines Protocol of the CCW should be strengthened to prevent the indiscriminate use of APLs, governments remained divided on a further course of action. Some countries championed the idea of an APL ban, but the Review Conference featured consensus decision-making, where the objections of a single state would have been enough to prevent the adoption of an APL ban. Thus, the conference participants chose to focus instead on bolstering the Mines Protocol, in order to reduce civilian casualties and the use of civilian lands as minefields.

The revised Protocol on Prohibitions or Restrictions on the Use of Mines, Booby-Traps and Other Devices, adopted by consensus at the final session of the Review Conference on May 3, 1996, had several improvements over the 1980 Mines Protocol (Matheson 1997). First, the scope of the 1980 Protocol was expanded to cover domestic as well as international conflicts. Moreover, it was recognized that some sections of the revised Protocol would also apply during peacetime, such as the provisions on the recording and monitoring of minefields, the transfer of mines, and consultations and compliance. Second, it was agreed that all APLs which are remotely deliverable should be equipped with a self-destruct (SD) device that would activate within thirty days of the mine's placement with an accuracy rate of ninety percent. The APLs would be outfitted with a backup self-deactivation (SDA) mechanism as well, which would be initiated within 120 days with a combined reliability of 99.9 percent. Third, the conference participants concurred that those APLs which are not remotely deliverable must either be fitted with SD and SDA devices or be confined to minefields that are protected by special measures to prevent civilian casualties, such as fencing and clearly visible markings. Fourth, the participants decided that all APLs should contain at least eight grams of iron in order to make them detectable by mine detection equipment.

Fifth, mines that were designed to be detonated by the operation of mine detection equipment were banned. Sixth, the Review Conference agreed that mines may not possess an antihandling device with a longer lifetime than the SDA mechanism. Seventh, it was accepted that the responsibility for mine maintenance and clearance lies with the party that deployed the mines. In addition, once a conflict has ended, mine clearance must begin without delay. Eighth, the Review Conference decided that annual meetings would be held to review the Protocol and discuss compliance with it. People who willfully kill or injure civilians by violating the Protocol would be prosecuted. Ninth, the revised Protocol proscribed the transfer of prohibited mines to any recipient, and banned the transfer of mines to states and nonstate actors who have not signed the revised Protocol or agreed to respect its provisions. Tenth, the requirements for recording the location of mines were made tougher. Eleventh, the conference added provisions for the protection of peacekeeping forces and humanitarian missions from APLs that were deployed in their areas of operation. Finally, the revised Protocol encouraged mutual assistance and technology transfer for demining activities, as well as to ensure compliance with the requirements of the Protocol.

Despite the considerable amount of changes to the 1980 Mines Protocol that were adopted by the Review Conference, advocates of a total ban on APLs, such as the ICRC and the International Campaign to Ban Landmines (ICBL), were disappointed (Maresca and Maslen 2000; Williams and Goose 1998). In the words of ICBL members Jody Williams and Stephen Goose, "from beginning to end, the preparatory sessions and the negotiations fell victim to an incremental approach that limited progress to adjustments within the existing framework of the [CCW]" (Williams and Goose 1998, 31). The ICRC was unsuccessful in its calls for a redefinition of "antipersonnel landmine" to include munitions that were originally designed for another purpose, but may be used as an APL (Elwell 1998; Maresca and Maslen 2000). Weapons that have the same effects as APLs, but are not classified as such, may escape the restrictions that were included in the revised Protocol, and thus cause more civilian casualties. Furthermore, the weaker, revised Protocol was adopted despite public opinion surveys in twenty-one countries which showed tremendous support for a total ban on APLs.

Nongovernmental organizations mobilize against landmines

The International Campaign to Ban Landmines was formed in October 1992, following a meeting in the New York office of Human Rights Watch. Six NGOs banded together and issued a joint call to ban APLs, thereby launching an international campaign. These NGOs, which became the steering committee of the ICBL, included Handicap International (France), Medico International (Germany), Mines Advisory Group (United Kingdom), Human Rights Watch (US), Physicians for Human Rights (US), and the Vietnam Veterans of America Foundation (VVAF, US). Jody Williams of the VVAF was appointed as the coordinator of the ICBL. Since May 1993, when the ICBL hosted the first ever NGO-sponsored international landmine conference, more than 1,200 NGOs in around sixty countries have joined the coalition. The ICBL emphasized two objectives. First, there was a need for a global ban on the use, production, stockpiling, and transfer of APLs. Second, the resources devoted to humanitarian demining and the assistance of landmine victims had to be increased (Williams and Goose 1998).

Key to the ICBL's success was the decentralized nature of its operations. The ICBL was not organized hierarchically, with a central office and bureaucracy. Instead, member organizations were free to pressure their own governments for an APL ban the way they saw fit. According to Jody Williams and Stephen Goose, "much of the unity and success of the coalition can be traced to a commitment to a constant exchange of information—both internally among members of the ICBL as well as with governments, the media, and the general public" (Williams and Goose 1998, 23). The ICBL members proved to be adept at cultivating close relationships with media outlets, which began to endorse the idea of a global APL ban. With the growing media campaign of shame, it became very difficult for militaries to justify publicly their need for APLs.

The ICBL also flourished through the development of personal relationships between its members, government agents, and military officials. International conferences sponsored by the ICBL, such as those held in Cambodia in 1995 and in Mozambique in 1997, presented opportunities for members to share information, attend training workshops, and develop plans for action at the regional and international levels. Moreover, the ICBL recognized that the first positive steps toward a ban would probably be taken in countries with a democratic political culture, where political activism by NGOs is permitted. Hence, the ICBL concentrated its campaign on North America, Europe, Australia, and New Zealand during the first few years. Once its network had become established in the North, the ICBL expanded its activities throughout Asia and Africa. Maxwell Cameron (2002) suggested that the ICBL felt more comfortable cooperating with the like-minded middle powers like Austria, Belgium, and Canada, rather than with the governments of major powers like France, Japan, and the United Kingdom. The partnership that was formed between the ICBL and the middle powers would be instrumental for achieving the APL ban.

Unilateral state action on landmines

In 1992, the United States became the first country to take unilateral action on the landmine issue. The year before, the Women's Commission for Refugee Women and Children had testified before the US Congress on the necessity of a landmine ban (Price 1998). Following consultations with ICBL coordinator Jody Williams, the VVAF, and other NGOs, Senator Patrick Leahy (D-Vermont) and Representative Lane Evans (D-Illinois) wrote legislation for a one-year moratorium on the export of APLs by the United States, which President George H. W. Bush signed into law in 1992 (Leahy 1997; Williams and Goose 1998). France then responded in February 1993, making its voluntary abstention from exporting APLs, in place since the mid-1980s, into official policy. Later that month, after experiencing pressure from Handicap International and the French anti-landmine campaign, the French government called for a Review Conference of the CCW. Shortly afterward, more than a dozen states announced export moratoria of their own.

In June 1994, under pressure from the Swedish anti-landmine campaign led by *Rädda Barnen* (Save the Children), the Swedish parliament voted for the government to work toward achieving a global ban on APLs. Sweden would later table an amendment to the Mines Protocol of the CCW that would have banned APLs, but it was not adopted due to insufficient support from other states at the time. On August 2, 1994, the Italian Senate ordered the government to ratify the Mines Protocol immediately, adopt a moratorium on the export of APLs, cease the production of APLs in Italy and by Italian companies abroad, and promote mine clearance in APL-infested countries. In the words of Jody Williams and Stephen Goose, "this was a critical move on the part of a country that was considered to be one of the three most significant producers and exporters of [APLs] in the world" (Williams and Goose 1998, 27).

In his address to the UN General Assembly on September 26, 1994, US President Bill Clinton called on the international community to eliminate APLs. The United States then sponsored a General Assembly resolution which urged states to adopt export moratoria, and also encouraged international cooperation to achieve the goal of eradicating APLs. But Jody Williams and Stephen Goose indicated that "the combination of Clinton's remarks and the resolution erroneously led many to believe that the US administration was finally following the lead on the issue shown in the US Congress and was signaling its willingness to move rapidly towards a ban" (Williams and Goose 1998, 27). Two countries did make hasty progress, however.

In March 1995, Belgium became the first state to ban the use, production, trade, and stockpiling of APLs, while Norway did the same three months later. Representatives from both governments have admitted that pressure from NGOs was the deciding factor to enact their bans.[6] By mid-1997, around thirty countries had unilaterally prohibited the use of APLs, twenty had banned production, fifteen had either begun or finished destroying their stockpiles, and more than fifty had made APL export illegal (Lenarcic 1998). But of the major powers which had announced their support for a comprehensive global ban, such as France, Germany, Japan, the United Kingdom, and the United States, only Germany had made a unilateral renunciation of the use of APLs (Price 1998).

The Ottawa Process

The Ottawa Process was born from the frustrations of the like-minded states, international organizations, and NGOs that the CD was unwilling to derive a total ban on APLs (Lawson et al. 1998). The EU, the OAU, the Organization of American States (OAS), UN Secretary-General Boutros Boutros-Ghali, UN High Commissioner for Refugees Sadako Ogata, and Pope John Paul II all called for an APL ban (Leguey-Feilleux 2009; Lenarcic 1998). In October 1996, the EU introduced a common moratorium on APL exports to all destinations.[7] A UN General Assembly resolution passed in December 1996 called on states to negotiate a legally binding international ban on the use, stockpiling, production, and transfer of APLs as quickly as possible.

In January 1996, the Dutch Campaign to Ban Landmines suggested that the ICBL should host a conference of pro-ban states (Leguey-Feilleux 2009). The ICBL was initially cold to this innovative idea of an NGO convening an intergovernmental conference, as they feared that momentum for the ban would be lost if states rejected their invitations. But due to the decentralized nature of the ICBL, the Dutch campaign was free to organize its own conference, which it promptly did. Twenty-two states were invited, of which seven attended (Austria, Belgium, Denmark, Ireland, Mexico, Norway, and Switzerland), as well as Canada, which had not been formally invited because the Canadian government had not yet endorsed a total APL ban. The success of this initial collaboration between NGOs and states spurred a second conference in Geneva in April 1996, where fourteen countries participated.

The Canadian government decided to exercise leadership on the landmine issue by co-hosting, together with the NGO Mines Action Canada, a conference on October 3–5, 1996, titled "Towards a Global Ban on Anti-Personnel Mines." Prior to the Ottawa conference, Canadian officials discussed with other pro-ban actors the issue of who to invite to the talks, since the participation of skeptical parties may have impeded progress at the conference (Lawson et al. 1998). A decision was made to invite states to participate on the basis of self-selection. A draft Final Declaration of the Ottawa conference was drawn up before the conference and circulated. Those states which were willing to support the Declaration were invited to attend as participants, while those who did not were welcomed as observers. International organizations and NGOs who supported an APL ban, such as the ICBL, the ICRC, and the United Nations Children's Fund (UNICEF), were invited to participate in the Ottawa conference. In total, fifty states who pledged support for the Ottawa Declaration attended the conference, as well as twenty-four observer countries and dozens of NGOs (Elwell 1998).

The fifty states who signed the Ottawa Declaration, including France, the United Kingdom, and the United States, made a commitment to cooperate in order to ensure that a legally binding international agreement banning APLs would come into force as soon as possible (Lenarcic 1998). An "Agenda for Action on Anti-Personnel Mines" was also adopted, which described a series of activities to be carried out by the conference participants in order to generate the political will for an APL ban (Lawson et al. 1998). But the most surprising event occurred on the last day of the conference. In his final speech, Canadian Foreign Minister Lloyd Axworthy invited the conference participants to work with Canada to negotiate and sign an APL ban treaty by December 1997, and furthermore, implement the treaty by the year 2000.[8] With the setting of a deadline for action on the landmine ban, the Ottawa Process was launched.

The Ottawa Process consisted of two tracks (Lenarcic 1998). Track one involved fast-track diplomatic negotiations on a ban treaty. Maxwell Cameron (2002) emphasized that the primary reason why the Ottawa Process would ultimately be successful was because it did not adopt the cumbersome, slow, consensus decision-making of the CD. Instead, a fast-track diplomatic approach was utilized, which would generate a treaty with few exemptions. This would be no easy task, as Robert Lawson and his co-authors indicated that "getting dozens of countries from all regions of the world to a single negotiating table to develop a ban convention in less than a year would require an almost unprecedented degree of diplomatic choreography" (Lawson et al. 1998, 166). Therefore, in order to achieve the objective of an APL ban, Canada worked closely with the other like-minded states who were members of the Ottawa Process core group. This group originated from a meeting in early 1996 between Austria, Belgium, Canada, Denmark, Ireland, Mexico, Norway, Switzerland, the ICBL, and the ICRC, to derive a strategy for achieving the APL ban that the CD was unwilling to address. In February 1997, the Ottawa Process core group met formally for the first time, and with the addition of Germany, the Netherlands, the Philippines, and South Africa, the group became more representative of the different regions of the world.[9]

Each of the like-minded states assisted the campaign in significant ways (Lawson et al. 1998; Lenarcic 1998). Discussions between Austria and Canada in early 1997 generated a draft plan for putting the diplomatic process into motion. Austria wrote a rough draft of an APL ban convention, which it presented at the Ottawa conference, and hosted an international meeting of landmine specialists from 111 states in Vienna in February 1997, in order to discuss the draft convention. In April 1997, a technical meeting of landmine experts from 120 countries was held in Bonn, Germany, to deliberate on the verification and compliance mechanisms that would be included in the ban treaty. Belgium hosted an APL conference in Brussels in June 1997, which was attended by 155 states. The conference ended with ninety-seven countries signing the Brussels Declaration, which called for a total ban on APLs, the destruction of APLs which had been stockpiled or removed, and international cooperation and assistance for the enormous task of mine clearance. Switzerland played host to several meetings of the core group in Geneva, while the formal negotiations on the APL ban convention were hosted by Norway in the fall of 1997.

The core group also promoted the idea of an APL ban at both the CD in Geneva and the UN in New York City. The sharing of information and close coordination between members made the core group more cohesive over time. Membership in the core group broadened some more as the Ottawa Process evolved, to eventually include Brazil, Colombia, France, Malaysia, New Zealand, Portugal, Slovenia, the United Kingdom, and Zimbabwe (Kongstad 1999). But in order to ensure that fast-track diplomacy would produce an effective APL ban, only like-minded states were invited to join the core group. As Maxwell Cameron explained:

> Since the clarity of the goal—a total ban on [APLs]—was essential to maintaining core group unity, when faced with the trade-off between increasing the number of supporters of a ban treaty and avoiding exceptions, the core group opted for a clean convention that would establish an unequivocal norm. (Cameron 2002, 81)

Following the February 1997 Vienna conference, Canada produced a paper detailing the procedures for formal negotiations on a ban treaty, which it presented at a meeting of the core group in Vienna in early March. During the meeting, Canada approached Norway about hosting a future conference on APLs, which Norway had expressed interest in doing at the October 1996 Ottawa conference. According to Robert Lawson and his colleagues, "the generosity and rapidity with which Norway responded to the enormous diplomatic and organizational challenge of hosting an international negotiation were key to the ultimate success of the Ottawa Process" (Lawson et al. 1998, 171). Since the Ottawa Process was receiving considerable criticism for the unorthodox way in which it was initiated outside normal diplomatic channels, the core group felt that the holding of a traditional diplomatic conference would help convince skeptical countries to join the APL ban campaign. The core group decided to invite Ambassador Jacob Selebi, a widely respected South African diplomat and senior official of the African National Congress, to chair the Oslo conference.

While the diplomatic negotiations of track one were underway, the Ottawa Process was simultaneously embarking on track two: the development of political support for an APL ban through the implementation of the Ottawa conference's "Agenda for Action on Anti-Personnel Mines" (Lawson et al. 1998; Lenarcic 1998). The momentum for a ban was generated through a series of regional conferences. The core group calculated that by getting the Southern, mine-infested states aboard the campaign, the Ottawa Process could avoid being stalled by a North-South split on issues related to the APL ban. The ICBL held the Fourth International NGO Conference on Landmines in Maputo, Mozambique on February 25–28, 1997. Four hundred and fifty NGO representatives from sixty countries attended the Maputo conference, where the ICBL representing more than eight hundred NGOs announced its support for the Ottawa Process, and South Africa declared a unilateral ban on APLs. Canada, South Africa, the OAU, the ICBL, and the ICRC then organized a pan-African landmine conference in Kempton Park, South Africa on May 19–22, 1997. By the end of the conference, forty-three out of the fifty-three OAU members had pledged their support for the Ottawa Process. That same month, Sweden hosted a meeting of governments and NGOs from Central and Eastern Europe. In June, Turkmenistan played host to the first APL ban conference ever held in Central Asia. This was followed in July by an ICBL-sponsored regional colloquium in Sydney, Australia, and a three-day seminar organized by the Philippines and the ICRC that was intended to generate more support for an APL ban in the Asia-Pacific region.

Around ten multilateral meetings at the global, regional, and subregional levels were held during the eleven-month period prior to the Oslo conference. The meetings were intended to pressure national decision-makers, through both state-led diplomacy and NGO-led advocacy, to adopt an APL ban. In regions such as Latin America and Asia, where NGOs were less capable of promoting the APL ban effectively, diplomats and political leaders from the core group countries made the rounds in order to convince governments to support the Ottawa Process, even if they were faced with opposition from military establishments which dismissed the idea of a ban. The anti-landmine campaign gained many important supporters, including Princess Diana, Archbishop Desmond Tutu, Jimmy Carter, Gracia Machel, Kofi Annan, and Queen Noor (Lawson et al. 1998; Manley 1998). The regional strategies of the Ottawa Process began to pay off when the Central American Common Market (CACM) and the Caribbean Community and Common Market (CARICOM) became the first regional organizations to announce their support for the Ottawa Process.

On September 1–18, 1997, the Diplomatic Conference on an International Total Ban on Anti-personnel Landmines met in Oslo (Lawson et al.

1998; Maresca and Maslen 2000). The mood was somber as Princess Diana, one of the most prominent advocates of an APL ban, was killed in a car crash in Paris the day before the conference opened. The conference attracted eighty-seven full participants and thirty-three observer states, and discussions focused on the third Austrian draft of the treaty. In contrast to the consensus decision-making procedure of the CD, the Oslo conference permitted decisions to be taken by two-thirds vote if consensus could not be reached. The first two days of the Oslo conference were devoted toward identifying issues of contention, which were then divided among the delegations from Austria, Brazil, Canada, Ireland, Mexico, and South Africa for more consultation and problem-solving.[10] The skilled leadership of the conference Chair, Ambassador Jacob Selebi, and the strong commitment to the ban treaty that was expressed by many governments ensured that no compromises were accepted that would have severely weakened the treaty. On the final day of the conference, the participants adopted the Convention on the Prohibition of the Use, Stockpiling, Production and Transfer of Anti-personnel Mines and on Their Destruction.

On December 3–4, 1997, 2,400 participants, including more than five hundred members of the international media, attended the second Ottawa landmines conference, titled "A Global Ban on Landmines: Treaty Signing Conference and Mine Action Forum" (Faulkner 1998; Lawson et al. 1998; Maresca and Maslen 2000). One hundred and twenty-two states signed the APL ban convention, and three countries—Canada, Ireland, and Mauritius—ratified it immediately. The conference featured twenty "Mine Action Roundtables," where the world's leading landmine experts discussed future mine action efforts. Their recommendations were published in the final report of the conference, *An Agenda for Mine Action*. Canada and the rest of the Ottawa Process core group used the conference as an opportunity to launch the "Ottawa Process II." This new phase of the anti-landmine campaign would involve the mobilization of national governments, international organizations, and NGOs, in order to achieve the objectives of deriving a global action plan to convince all states to sign the treaty, clearing the millions of mines remaining in the ground, and providing assistance to landmine victims (Lenarcic 1998). The participating states pledged more than $500 million for mine action programs globally.

The key members of the Ottawa Process coalition attended the one-day "Ottawa Process Forum" immediately after the conference ended, where they examined the lessons learned from the campaign (Lenarcic 1998). In addition, on December 6–7, Mines Action Canada hosted a two-day seminar where NGO members could consult and plan for the Ottawa Process II. The success of the Ottawa Process was underscored by the awarding of the Nobel Peace Prize to Jody Williams and the ICBL in Oslo on December 10, 1997, just a few days after the signature of the Ottawa Convention.

The United States and the Ottawa Process

Although the United States was an early leader in the campaign to ban landmines and implemented the world's first landmine export moratorium, it refused to support the Ottawa Process, because the proposed ban treaty did not include exemptions for American APLs. In the spring of 1996, the United States conducted an internal policy review to determine the military utility of landmines (Kirkey 2001). The results were made public in May 1996. According to the Public Affairs Office of the US Department of Defense, the United States required APLs for the protection of American forces in Korea and Guantanamo Bay, Cuba, as well as for training exercises (Matthew and Rutherford 1999). APLs were regarded as particularly useful for enhancing the effectiveness of antitank landmines (ATLs).

On May 16, 1996, President Bill Clinton clarified the landmine policy of the United States (Kirkey 2001). First, the United States was committed to the adoption of an international treaty that would eliminate all landmines. Second, the United States intended to dispose of all landmines that were not self-detonating or self-deactivating (i.e. "dumb mines"), with the exception of the more than one million landmines being used to protect American and South Korean military forces from a North Korean attack. Finally, the United States would continue to use self-detonating or self-deactivating landmines (i.e. "smart mines") until either effective alternatives to replace them would be derived or an international landmine ban treaty would come into force. The objective was to eliminate APLs by 2006–10 (Matthew and Rutherford 1999).

The United States was also wary of the fast-track diplomacy of the Ottawa Process. The United States attended the first Ottawa conference in October 1996, but believed that Canadian Foreign Minister Lloyd Axworthy's appeal for a ban treaty to be negotiated and signed within one year was an unrealistic goal (Lenarcic 1998). On November 4, 1996, the United States introduced a resolution in the UN General Assembly, originally drafted by Canada and co-sponsored by eighty-four states, calling on countries to derive a comprehensive APL ban treaty as soon as possible. The resolution passed by a vote of 156-0 on December 10, 1996, but ten major users and producers of APLs abstained from voting.[11] These ten states criticized the resolution for not recognizing that APLs have a legitimate role to play in the defense policy of a state, not discussing alternatives to APLs, and not considering the use of APLs by terrorists. The pro-APL states voted in favor of another UN resolution calling for the strengthening of the CCW, which was adopted by consensus.

In January 1997, the United States announced that while it welcomed the efforts of the Ottawa Process, it had made the decision to begin negotiations on an APL ban treaty within the CD. The United States preferred to launch the initiative in this forum for two reasons: the CD was considering the adoption of a more holistic arms control approach, and China and Russia were both members of the CD (Williamson 2000). The United States believed that the major producers of APLs would not participate in the Ottawa Process, hence, the CD would be the appropriate forum for discussing the landmine issue with the Chinese, Russians, Indians, and Pakistanis (Lenarcic 1998; Price 1998).[12] Faced with strong domestic pressures from the US Campaign to Ban Landmines (USCBL), the Clinton administration also decided to turn the 1992 export moratorium into a permanent ban in January 1997 and capped the American stockpile of APLs at its existing level, which was later discovered to be around fourteen million mines (Wareham 1998). Since 1997, the United States has not produced any antipersonnel landmines (ICBL 2006).

But in February 1997, the CD adopted an agenda which did not include APLs (Lenarcic 1998; Wareham 1998). Throughout 1997, the United States and other states attempted to place landmines on the CD agenda, but with no success. Despite American insistence that the CD delve promptly into the issue of landmines, a proposal to create an ad hoc committee on landmines within the CD was blocked by nonaligned states who wanted to discuss nuclear disarmament first. Some states opposed the placement of an APL ban on the CD agenda because they wanted to avoid jeopardizing the Ottawa Process, while other states argued that precedence should be given to implementation of the revised Mines Protocol, and to ongoing negotiations on the CCW, before discussing an APL ban in the CD. Frustrated with the lack of progress in the CD, the United States declared that it would pursue other channels if the CD did not place an APL ban on its agenda by the end of June 1997.[13]

In July 1997, the Landmine Elimination Act was introduced in both Houses of the United States Congress, with fifty-nine senators and 190 representatives as co-sponsors (Wareham 1998). The bill, which banned new American deployments of APLs after January 1, 2000, never came to a vote, as Senator Patrick Leahy (D-Vermont) and Senator Charles Hagel (R-Nebraska) withheld action on the bill, in order to give the Clinton administration an opportunity to participate in the Ottawa Process. A letter signed by 164 members of the House of Representatives also indicated that the Congress was backing American participation in the Ottawa Process. There was significant domestic opposition to both the Landmine Elimination Act and the Ottawa Process, however, particularly from the Pentagon, the Joint Chiefs of Staff, and Senator Jesse Helms (R-North Carolina), who served as the Chairman of the Senate Armed Services Committee.

Nevertheless, the Clinton administration reversed its stance in August 1997, and announced that the United States would join the Ottawa Process. The United States signed the Brussels Declaration, a prerequisite for attending the Oslo conference. Although the United States claimed that other states would follow the American lead and attend the conference, Japan and Poland proved to be the only significant countries to follow suit. At Oslo, the American delegation proposed critical, nonnegotiable modifications to the treaty that would have weakened it considerably had the revisions been accepted. The United States demands included an exemption for the continued use of APLs in Korea; a redefinition of APLs so that the United States could keep its dual antitank and antipersonnel landmine systems; a tougher treaty ratification process and a nine-year deferral period for compliance with certain provisions; stronger verification procedures; and an option for a state to withdraw from the treaty if it perceives that its supreme national interests are threatened (Kirkey 2001; Wareham 1998).

With the exception of the stronger verification measures, however, the United States failed to get its proposals included in the treaty. Hence, the United States refused to sign the Ottawa Convention in December 1997. The Clinton administration did adopt some unilateral initiatives, nevertheless. The administration announced that the United States would develop APL alternatives that would end American reliance on both self-destruct APLs by 2003 and its mines in Korea by 2006. Moreover, American funding for mine clearance programs would be increased by twenty-five percent beginning in 1998 (Lenarcic 1998). In a May 15, 1998, letter to Senator Patrick Leahy, Samuel Berger, the Assistant to the President for National Security Affairs, stated that the United States would sign the Ottawa Convention by 2006, if suitable alternatives to American APLs and mixed antitank systems would be derived by then (Kirkey 2001). President Clinton made this timetable official with the Presidential Decision Directive Number 64 of June 23, 1998.

But in November 2001, the Department of Defense recommended that the United States should both abandon its commitment to join the Ottawa Convention and discard some parts of the American program to develop alternatives to APLs (ICBL 2002). On February 29, 2004, the George W. Bush administration unveiled its landmine policy. Although the Bush administration announced a fifty percent increase in spending on mine action programs in 2005, the administration's decisions to continue using self-destructing landmines indefinitely, to extend the use of long-lived landmines until 2010, and to break President Clinton's promise to sign the Ottawa Convention by 2006, were condemned by the ICBL (Wixley 2004).

In November 2009, the Barack Obama administration revealed that after conducting a review of the American landmines policy, it had decided not to join the Ottawa Convention (HRW 2009). But despite American nonparticipation in the APL ban, the United States has adopted its own anti-APL policies, including the nonuse of APLs since 1991, a ban on APL exports since 1992, and a halt on the production of APLs since 1997. The US government has no plans to procure APLs in the future.

The results of the Ottawa Process

With the fortieth ratification of the Ottawa Convention by Burkina Faso in September 1998, the treaty entered into force in March 1999 ("World Watch: Ouagadougou" 1998). As of September 2011, 156 states had either signed or acceded to the treaty, and 154 had ratified it (ICBL 2011b). The rapid impact of the Ottawa Convention has been astounding. The number of known producers of APLs fell dramatically by 2001, from fifty-four to fourteen states ("Curbing Horror; Landmines" 2001). By 2010, only twelve states either manufactured APLs or reserved the right to do so (ICBL 2010b).[14] Remarkably, the global trade in APLs has been slowed to a trickle, reduced to a small number of illegal and undisclosed transfers. It is incredible to note as well that at least fifty-nine nonstate armed groups in thirteen countries have also pledged to halt their use of APLs (ICBL 2009b).

Each party to the Ottawa Convention is required to destroy all of their stockpiles of APLs no later than four years after the entry into force of the treaty for the country, and remove all APLs from their territory within ten years. By 2010, eighty-six states parties had completed the destruction of their stockpiles, destroying over forty-five million APLs (ICBL 2010b).

But on the negative side, Human Rights Watch reported in 2003 that by then only thirty-five states parties had passed domestic laws to prevent, suppress, or punish activities prohibited by the treaty (HRW 2003). Furthermore, as of 2008, seventy-one states parties had exercised the option, under Article 3 of the Ottawa Convention, of retaining some APLs for training and development purposes (ICBL 2009a). More than 197,000 APLs have been retained under this exception, but in some cases they have not been used for research purposes. Unfortunately, few states parties have provided explanations for their decisions to retain APLs that they are not using. Turkey (15,125 APLs), Bangladesh (12,500), and Brazil (10,986) accounted for nearly twenty percent of all mines retained under Article 3. Sweden (7,364 APLs), Greece (7,224), Australia (6,785), Algeria (6,090), Croatia (6,038), and Belarus (6,030) are other leading APL retainers among states parties. Thirty-eight states parties each retained between one thousand and five thousand APLs, while twenty-four states parties each retained fewer than one thousand APLs.

Additionally, as of September 2011, thirty-eight states had still not signed the Ottawa Convention (ICBL 2011a). Included on this list are China, Russia, the United States, nuclear rivals India and Pakistan, perennial adversaries North and South Korea, and several Middle Eastern states, including Egypt, Iran, Israel, Lebanon, Saudi Arabia, Syria, and the United Arab Emirates. States that are not parties to the Ottawa Convention have stockpiled more than 160 million APLs, with the vast majority held by China (an estimated stockpile of 110 million mines), Russia (24.5 million), the United States (10.4 million), Pakistan (estimated at six million), and India (estimated at between four and five million) (ICBL 2009a). Furthermore, a few of the states parties who pledged to destroy their APL stockpiles have failed to deliver on their promises. In 2010, four states parties—Belarus, Greece, Turkey, and Ukraine—were in serious violation of the Ottawa Convention for having missed their stockpile destruction deadlines (ICBL 2010b).

Nevertheless, it is unmistakable that the APL ban has had a remarkable impact. In the period 1999–2000, shortly after the Ottawa Convention entered into force, eleven governments began new use of APLs in twenty conflicts, and at least thirty rebel groups deployed APLs (Kingman 2000). But by 2009–10, Myanmar's armed forces were the sole national military to use APLs, while nonstate armed groups used APLs in only six countries (ICBL 2010b).

Since most of the world's landmines are possessed by countries which are not parties to the APL ban treaty, the problem of APLs will not be resolved until all landmine producers and users have signed and ratified the Ottawa Convention, and have implemented its provisions. The Ottawa Process can be applauded, however, for the considerable success it has had in moving the world much closer to the point where APLs may become history, especially when compared to the lack of progress in the CD. In addition, the countries which have had the majority of landmine casualties have signed the Ottawa Convention; hence, the most mine-contaminated states are covered by the treaty (Price 1998). Most important, the Ottawa Process has succeeded in generating a new international norm that stigmatizes the use of APLs. As more states sign the Ottawa Convention, greater pressure to emulate is placed on the remaining holdout countries. Many of the states not parties have taken measures to comply with the Ottawa Convention, even if they have not acceded to the APL ban treaty (ICBL 2006). Since 1997, the UN General Assembly has passed an annual resolution calling for the universalization of the Ottawa Convention. In December 2008, 163 states voted in favor, none opposed, and eighteen abstained (ICBL 2009a). What is remarkable is that eighteen states not parties to the Ottawa Convention voted in favor of the resolution, including China.

From November 29 to December 3, 2004, the Nairobi Summit on a Mine-Free World was held, fulfilling the UN's obligation to hold a review conference on the Ottawa Convention five years after it came into effect (Geneva International Centre for Humanitarian Demining 2005). The summit featured the participation of 135 states, including twenty-five states not parties to the Ottawa Convention, and more than 350 representatives of NGOs. Ambassador Wolfgang Petritsch of Austria served as President of the Nairobi Summit. The participants conducted a comprehensive review of the results of the APL ban treaty and adopted the *Nairobi Action Plan 2005–2009*, which outlined seventy specific actions that states parties would undertake in the five-year period following the summit. The states parties agreed to hold annual formal and informal meetings in preparation for a second review conference in 2009. In addition, the states parties adopted *Towards a mine-free world: the 2004 Nairobi Declaration*, which reiterated their commitment to rid the world of APLs and protect people from becoming victims of these indiscriminate weapons.

The Second Review Conference of the Mine Ban Treaty was held in Cartagena, Colombia, from November 29 to December 4, 2009 (ICBL 2010a).

A total of 108 states parties to the Ottawa Convention participated, as did twenty states not parties. The ICBL contingent included 419 representatives from seventy-three countries, the largest delegation the ICBL had ever sent to a landmines conference. For the first time ever, the United States participated as an observer at a meeting on the APL ban treaty. With the Norwegian Ambassador Susan Eckey serving as President of the Second Review Conference, the summit participants adopted the *Cartagena Declaration* reaffirming their commitment to the APL ban, and issued the *Review of the Operation and Status of the Convention 2005–2009*. They also agreed to the *Cartagena Action Plan 2010–2014*, which emphasized assistance to APL victims, mine clearance, the destruction of APL stockpiles, education on the risks posed by APLs, and the necessity of further international cooperation. In addition, Albania, Greece, Rwanda, and Zambia announced that they had finished clearing all their APLs, bringing the total number of mine-cleared states to sixteen. The states parties also promised to continue holding meetings on a regular basis.

Conclusion

The case of the APL ban illustrates Finnemore and Sikkink's (1998) three stages of norm evolution. During the first stage of "norm emergence," the middle powers exercised skilled leadership as norm entrepreneurs. Starting with Canadian Foreign Minister Lloyd Axworthy's bold call for the quick realization of a ban on APLs within fourteen months, the Ottawa Process core group employed fast-track diplomacy in order to ensure that an effective ban treaty with few exemptions would be produced. The core group drafted a treaty that would ban the use, stockpiling, production, and transfer of APLs, and collaborated with NGOs in order to persuade other states to join the campaign by emphasizing the humanitarian toll of APLs. By December 1997, a "tipping point" had been reached when the middle powers had generated sufficient political will for the second stage of "norm cascade" to occur, and 122 states proceeded to sign the Ottawa Convention. The global APL ban is currently in the third stage of "internalization," as states have been developing respect for the new norm and have been modifying their behavior to comply with it.

The United States was an early leader in the campaign to ban APLs, but refused to support the Ottawa Process for two reasons. First, Washington objected to the fast-track diplomatic strategy used by the core group. The United States preferred to rely on the consensus decision-making of the CD, but was disappointed when the CD took no action on APLs. Second, the United States disapproved of the draft treaty that was negotiated by the core group, because there were no exemptions in the treaty for the American landmines in South Korea. The core group was steadfast in resisting US pressures for the inclusion of exemptions that would have favored American interests, but would have also weakened the Ottawa Convention. But despite its unwillingness to sign the convention, the United States acquiesced to the APL ban, because it only posed a challenge to American security interests abroad, and did not threaten the core US national interest, which is the security of the American territory, institutions, and citizenry.

Notes

1 Hughes Aircraft, a subsidiary of General Motors, was the second largest landmine manufacturer in the United States after Alliant Techsystems. See Capellaro and Cusac 1997.
2 For a discussion of the landmine problem in Angola, see Winslow 1997. For an analysis of the landmine situation in Mozambique, see the Human Rights Watch Arms Project and Human Rights Watch/Africa 1994. To read about the difficulties of demining in Mozambique, see Purves 2001.
3 Paul Davies and Nic Dunlop (1994) provided an illustrated description of the impact of landmine warfare on communities in Cambodia.
4 The full titles of the Geneva Conventions, with their dates of signature in parentheses, are Convention Relative to the Treatment of Prisoners of War (27 July 1949); Convention (I) for the Amelioration of the Condition of the Wounded and Sick in Armed Forces in the Field (12 August 1949); Convention (II) for the Amelioration of the Condition of the Wounded, Sick and Shipwrecked Members of the Armed Forces at Sea (12 August 1949); Convention (III) Relative to the Treatment of Prisoners of War (12 August 1949); and Convention (IV) Relative to the Protection of Civilian Persons in Time of War (12 August 1949). The full titles of the two Additional Protocols, with their dates of signature in parentheses, are Protocol Additional to the Geneva Conventions of 12 August 1949, and Relating to the Protection of the Victims of International Armed Conflicts (Protocol I, adopted on 1 June 1977), and Protocol Additional to the Geneva Conventions of 12 August 1949, and Relating to the Protection of the Victims of Non-International Armed Conflicts (Protocol II, adopted on 1 June 1977).
5 The full title of the convention is the United Nations Convention on Prohibitions or Restrictions on the Use of Certain Conventional Weapons Which May be Deemed to be Excessively Injurious or to Have Indiscriminate Effects. It is also known as the United Nations Convention on Inhumane Weapons.
6 For a discussion on how NGOs influenced the government of Norway to pursue an APL ban, see Neumann 2002. To read about the roles of NGOs and other middle powers in convincing the Australian government to support the Ottawa Process, see Maley 2002.
7 Some regional organizations went so far as to declare themselves "mine-free zones." This was done by the CACM in September 1996, the CARICOM in December 1996, the OAS in 1996 and 1997, and the OAU in May 1997. See Lenarcic 1998.

8 Apparently, Lloyd Axworthy's bold speech setting a deadline for action on an APL ban even caught Canadian Prime Minister Jean Chrétien by surprise.
9 Not all of the middle powers were enthusiastic supporters of the Ottawa Process. Australia preferred the more inclusive diplomatic process of the CD, even though its objectives were less extensive. In fact, Australian officials were upset with Canadian Foreign Minister Lloyd Axworthy's call for states to negotiate a ban treaty within one year's time, as it took them by surprise. See Maley 2002.
10 These six middle powers held influential positions at the Oslo conference. The conference Chair, Ambassador Jacob Selebi, was from South Africa, while the other five states were the "Friends of the Chair," according to Robert Lawson et al. (1998, 177).
11 The ten countries were Belarus, China, Cuba, Israel, North Korea, Pakistan, Russia, South Korea, Syria, and Turkey.
12 Australia, France, and the United Kingdom also promoted the CD process initially. In fact, France was uncomfortable with the involvement of NGOs in the Ottawa Process. It took changes of government in each of these states before they hopped onto the Ottawa Process bandwagon. See Elwell 1998 and Price 1998.
13 The CD did adopt an Australian proposal and appointed a special coordinator on landmines in late June, which met the approval of the United States as well as France, Germany, and the United Kingdom. But progress on the landmine issue remained slow. See Lenarcic 1998.
14 The twelve states who reserved the right to manufacture APLs were China, Cuba, India, Iran, Myanmar, North Korea, Pakistan, Russia, Singapore, South Korea, the United States, and Vietnam. The *Landmine Monitor* claimed that only India, Myanmar, and Pakistan were actually manufacturing APLs. See ICBL 2010b.

CHAPTER FIVE

Establishing the International Criminal Court

On July 1, 2002, the Rome Statute of the International Criminal Court (ICC) entered into force. The creation of the ICC fulfilled a decades-old dream of establishing a permanent mechanism for trying individuals who are accused of mass atrocities. At the forefront of this successful human security initiative were the like-minded middle powers, whose skilled leadership was instrumental for ensuring the adoption of an effective mechanism that could achieve justice for genocide, crimes against humanity, war crimes, and the crime of aggression.

The fact that the ICC initiative attained its objectives is remarkable, considering that it encountered a strong opposition from the United States. As the hegemon in the contemporary international system, the United States frequently needs to engage in military operations abroad. Washington was concerned that US military personnel serving overseas would become targets for politically motivated prosecutions by the ICC. Furthermore, the United States doubted that the ICC would grant American defendants their rights to a jury trial and due process, which are protected under the Fifth and Sixth Amendments to the US Constitution.

In this chapter, I commence with a discussion of how the idea of an international criminal court evolved historically. The ensuing section describes the ICC that was created. I then turn to an analysis of how the like-minded middle powers exercised leadership on the ICC initiative. This is followed by an investigation of why the United States objected to the Rome Statute, and how Washington tried to weaken the treaty. I conclude with an exploration of the ICC in action.

An International Criminal Court: the history of the concept

The origins of international criminal tribunals

According to Reza Islami Some'a (1994), the first ever international tribunal was held in Breisach, Germany in 1474. The governor of Breisach, Peter Von Hagenbach, was convicted by twenty-seven judges of the Holy Roman Empire for permitting his soldiers to rape, murder, and steal property from the innocent civilians of Breisach. During the nineteenth century, treaties between Great Britain and other states led to the formation of international tribunals which were empowered to order the confiscation or destruction of ships that were engaged in the slave trade. The crews of these ships were not tried by the tribunals, but were returned to their home countries for punishment under domestic law.

The concept of a standing, international, adjudicating institution was born during the First Hague Peace Conference in 1899. Article 2 of the First Hague Convention for the Pacific Settlement of International Disputes required signatory states to use the good offices or mediation of a third party before resorting to conflict. Article 9 authorized commissions of inquiry that could investigate facts. The conference participants also agreed to set up a "permanent" court of arbitration, but only created a list of nonprofessional people who would sit as a court if and when the parties to a dispute requested their intervention (Islami Some'a 1994).

The Second Hague Peace Conference in 1907 produced the Hague Convention (XII) Relative to the Creation of an International Prize Court, which was concerned with the wartime practice of seizing ships and cargoes as prizes of war. The convention, which was signed but never entered into force, stipulated that the rulings of national prize courts on disputes related to the capture of property from neutral states or innocent civilian owners could be appealed to the International Prize Court (Islami Some'a 1994).[1] The Hague Convention (IV) Respecting the Laws and Customs of War on Land enunciated the rules of war for the states parties to the convention, and required that violators of the convention provide compensation to the injured parties (Sadat 2000). The Martens Clause to the Hague Convention IV stated that in cases where one or more of the belligerents are not parties to the convention, these countries and their inhabitants are protected by the "principles of the law of nations, as they result from the usages established among civilized peoples, from the laws of humanity, and the dictates of the public conscience" (Sadat 2000, 33). But the idea of criminal prosecution for those who violate the convention was not considered at the Hague Conference.

The aftermath of World War I

The next attempt to establish an International Criminal Tribunal occurred following the massive slaughter of World War I. Despite objections from the United States the 1919 Commission on the Responsibility of the Authors of the War and on the Enforcement of Penalties for Violation of the Laws and Customs of War proposed that an international high tribunal be created to hold trials for all enemy persons accused of violating the laws of war and humanity. Articles 227, 228, and 229 of the Treaty of Versailles authorized a special tribunal to try Kaiser Wilhelm II of Germany for the "supreme offence against international morality and the sanctity of treaties" (Sadat 2000, 34). But the trial was never held, as the neutral Netherlands refused to extradite the Kaiser after he had sought refuge there. The Allies refused to push for the prosecution of the Kaiser and others accused of war crimes, due to a fear that the trials would either provoke an armed revolt in Germany or start another war between Germany and the Allies (Islami Some'a 1994). German officers were prosecuted instead by the German Supreme Court in Leipzig. The trials were criticized by German citizens, since no Allied personnel who committed war crimes were prosecuted.

The Commission also recommended the prosecution of Turkish officials who carried out the Armenian genocide (1915–23), which killed around one and a half million Armenians in the Ottoman Empire. For the first time, the concept of "crime against humanity" was given a legal backing (Islami Some'a 1994). But since crimes against humanity did not exist under positive international law at that time, the Commission's report failed to designate such crimes for prosecution by an ICC. The Treaty of Sevres (1923), which called for Turkish prosecutions, was never ratified, and it was substituted by the Treaty of Lausanne (1927), which gave amnesty to the Turks. Hence, the Allies failed to seek justice for the first genocide of the twentieth century.

The first interwar discussion on the necessity of an ICC occurred in the League of Nations (Ferencz 1980; Von Hebel 1999). From June 16 until July 24, 1920, the Advisory Committee of Jurists met to draft a Statute for a Permanent Court of International Justice (PCIJ). The Committee also examined a proposal for creating a High Court of International Justice that would try crimes against the "international public order and the universal law of nations," but ultimately rejected this issue as being outside of the Committee's mandate (Von Hebel 1999, 17). In the end, the Legal Committee of the League did not accept the Jurists' recommendation that the court have compulsory jurisdiction, and gave member states the option of choosing whether to accept the decisions of the court, as well as the freedom to determine the degree of their compliance. In the words of Benjamin Ferencz:

> The failure of the League to accept an International Court with compulsory jurisdiction over those disputes which might lead to war meant that it was doomed to be a Court with limited authority, power or influence. . . . The new edifice for international society was being built on pillars made of sand. (Ferencz 1980, 36)

The issue of an ICC was discussed in various forums throughout the interwar period (Von Hebel 1999). In 1925, the Inter-Parliamentary Union declared that violations of the international order and the law of states should be defined, and that a chamber of the PCIJ should exercise jurisdiction over such offenses. The following year, the International Association of Penal Law made a similar argument. The International Law Association examined the subject of an ICC at three conferences, and decided at its 1926 Vienna meeting that the creation of an ICC was practical and feasible. Following the assassination of King Alexander of Yugoslavia and the French Foreign Minister in Marseilles on October 9, 1934, France introduced an initiative in the League of Nations to derive both an International Terrorism Convention and an ICC. Although the Diplomatic Conference on the Repression of Terrorism adopted both conventions and opened them for signature in November 1937, the Convention for the Prevention and Punishment of Terrorism was ratified solely by India, whereas the Convention for the Creation of an International Criminal Court was not ratified by any states.

World War II and the International Military Tribunals

The unbelievable atrocities of World War II sparked a new interest in the issue of establishing an ICC (Sadat 2000). But political pressures resulted in the postwar creation of the Nuremberg and Tokyo tribunals rather than the formation of an ICC as proposed by jurists. The International Military Tribunal (IMT) for the Far East has been criticized for its unfair treatment of many defendants, and has been mentioned as a model for "what a credible international criminal justice system ought not to look like" (Sadat 2000, 34). In contrast, the earlier IMT at Nuremberg made some major contributions to international criminal law. First, the Tribunal rejected the defendants' arguments that were based on state sovereignty and emphasized that individuals could be held criminally responsible under international law. Heads of state and individuals acting under orders were deemed to not be exempt from criminal charges. Second, the Nuremberg IMT stressed that the international duties of individuals transcend their obligations to obey the national laws of a state. Thus, international law has primacy over national law. Third, the Tribunal established that aggression is a crime, by ruling that individuals may be liable for both initiating a war, and the methods used for conducting the war. As a result, the act of war and transgressions against the laws of war were criminalized.

The United Nations and the International Law Commission

On December 11, 1946, the UN General Assembly adopted three resolutions: 1/94, 1/95, and 1/96. The first resolution created the Committee on the Progressive Development of International Law and its Codification. The second resolution issued a mandate for the Committee to prioritize the formulation of an International Criminal Code, which would be based on the principles recognized in the charter of the IMT at Nuremberg, as well as in its judgments. The third resolution emphasized that genocide was a crime under international law, and called on the UN Economic and Social Council (ECOSOC) to begin deriving a draft convention on the crime of genocide. The General Assembly adopted the Genocide Convention less than two years later (Von Hebel 1999).

The Committee discussed the idea of a permanent ICC, but was unsure whether this issue fell within the Committee's mandate. In its report to the General Assembly, the Committee suggested that the formation of an ICC may be desirable. The General Assembly debated the ICC issue, but only reached the conclusion that persons charged with genocide may be tried by either a tribunal in the state where the act was committed or an international penal tribunal whose jurisdiction has been accepted by all states involved (Von Hebel 1999).

In 1949, the International Law Commission (ILC) held its first ever meeting and discussed the codification of the Nuremberg principles as well as the creation of an ICC (Von Hebel 1999). The ILC appointed two Rapporteurs to analyze the ICC issue. The following year, the Rapporteurs issued conflicting recommendations, one in favor of an ICC and the other arguing that the time was not yet right for an ICC. Since a majority within the ILC wanted to establish an ICC, they created a Committee on International Criminal Jurisdiction. The Committee submitted a draft statute to the General Assembly in 1951, which was then referred to the member states for observations. In 1952, the General Assembly reviewed the draft statute as well as the feedback from the member states. Opinions on the necessity of an ICC varied considerably, therefore, the General Assembly created two more committees: one responsible for writing a new draft statute and the other for developing a definition of aggression. The Committee on International Criminal Jurisdiction presented a report in 1953, but the General Assembly decided to pass Resolution 9/898 on December 14, 1954, delaying discussion on an ICC until the definition of aggression had been clarified, and a draft Code of Offences had been derived.[2] The political consensus on the necessity of an ICC had not yet developed.

The initiative to create an ICC stalled in the UN over the next thirty-five years (Sadat 2000; Von Hebel 1999). The Cold War resulted in a period of inertia. It took twenty years for the General Assembly to adopt, by consensus, a definition of aggression, which it finally did with Resolution 29/3314 on December 14, 1974. But the definition has been criticized for not being exhaustive, for not binding the Security Council, and for permitting the Security Council to consider acts unmentioned in the definition as acts of aggression (Von Hebel 1999). In 1973, General Assembly Resolution 28/3068 adopted the International Convention on the Suppression and Punishment of the Crime of Apartheid, which allowed for the possibility of trial by an international penal tribunal. On February 26, 1980, the UN Commission on Human Rights requested, through Resolution 36/12, that an Ad Hoc Working Group study the feasibility of establishing an international criminal jurisdiction. But although the study was conducted, there was still insufficient political will to generate momentum for the ICC initiative. Indeed, a decade passed from the proclamation of General Assembly Resolution 36/106 in 1981—which requested that the ILC resume its work on a draft Code of Offences—until the ILC adopted the Code of Crimes against the Peace and Security of Mankind in 1991.

It took the thawing of East-West relations with the end of the Cold War for significant progress to be made toward an ICC (Sadat 2000; Von Hebel 1999). In 1989, Trinidad and Tobago sponsored General Assembly Resolution 44/39, requesting that the ILC place the ICC issue on the agenda of its next session. The ILC examined the issue only briefly during its 1990 session, but reached the conclusion that there was significant support for a permanent international criminal court. In 1992, the ILC created a Working Group to analyze the issue of establishing an ICC. The Working Group's report outlined the necessary conditions for the development of an ICC, but the group's consensus decision-making was criticized by some ILC members for generating lowest common denominator recommendations. The Working Group declared that it had completed its analysis of the feasibility of an ICC and concluded that a renewed mandate from the General Assembly was required before negotiations on a draft ICC statute could commence. Although the General Assembly was divided with regards to the necessity and feasibility of an ICC, on November 25, 1992, the Assembly responded to the ILC's request for a clear mandate by adopting Resolution 47/33, asking the ILC to prioritize the drafting of an ICC statute.

The post-Cold War international criminal tribunals

The post-Cold War era saw a return to the use of international criminal tribunals to prosecute war crimes, nearly forty-five years after the Nuremberg IMT. UN Security Council Resolution 827 of May 25, 1993, established the

UN International Criminal Tribunal for the former Yugoslavia (ICTY) in The Hague, with the aim of prosecuting war crimes, crimes against humanity, and acts of genocide that occurred in the former Yugoslavia since the start of the war in 1991 (Colwill 1995). The tribunal consisted of eleven judges, supported by a staff of around three hundred persons (Morton 2000). The first plenary session was held in The Hague in November 1993, and the tribunal began operating as a judicial body exactly two years later. The ICTY set precedence in that sexual assaults, categorized under the general heading of torture and enslavement, were investigated for prosecution as a crime against humanity for the first time (Cordner and McKelvie 1998). But the ICTY could only try individuals brought before it, which differed from the Nuremberg IMT, where trials in absentia were permitted. Furthermore, the ICTY was plagued by acts of noncompliance and defiance, particularly by the Yugoslav government of Slobodan Milosevic, which violated UN Security Council resolutions by delaying the issuing of visas for court investigators and refusing to hand over documentation or carry out search warrants (Pisik 1998).

Beginning on the night of April 6, 1994, the Rwandan genocide claimed the lives of between five hundred thousand and one million people in less than two hundred days (Goldstone 2000). In July 1994, the Security Council adopted Resolution 935, which created a Commission of Experts to investigate the violation of human rights in Rwanda (Morton 2000). The Security Council then established the UN International Criminal Tribunal for Rwanda (ICTR) in Arusha, Tanzania in November 1994. Modeled after the ICTY, the ICTR shared the same chief prosecutor, Louise Arbour from Canada. At least three hundred people were tried by the ICTR on charges of genocide, and more than one hundred were convicted and sentenced to death. The ICTR was more successful than the ICTY in terms of prosecuting offenders, but less active than the Rwandan national courts.

Human rights groups have applauded the creation of the ICTY and the ICTR for several reasons (Morton 2000). First, the investigations that were conducted prior to the establishment of the tribunals compiled a historical record of the events in each conflict. Second, there is a possibility that the investigations of war crimes may help moderate ethnic tension in these regions. By demonstrating that certain individuals are guilty of war crimes, the tribunals could educate the populations that solely the perpetrators are to be blamed, not entire ethnic groups. Third, the indictments handed down by the tribunals will help punish the guilty individuals by turning them into political pariahs. But the two tribunals' efforts at prosecuting accused war criminals revealed a weakness with the process regarding jurisdiction over suspects who are not in the custody of the court. In contrast to the Nuremberg IMT, which took place following the total defeat of Germany by the Allied Powers, the ICTY and the ICTR were set up in contexts where the conflicts were in a relative stalemate. Consequently, the accused war criminals were difficult to find and arrest, because they received shelter and support from their own ethnic groups. In the words of Jeffrey Morton:

> The inability of both tribunals to effectively arrest those indicted for genocide and war crimes, certainly most profound in the tribunal for the former Yugoslavia, undermines public confidence in the legal proceedings and keeps the possibility of future atrocities alive. (Morton 2000, 62)

Building an ICC: the path toward Rome

Drafting a statute

The creation of the ICTY and the ICTR gave a considerable boost to the campaign to establish an international criminal court. Sufficient political will for setting up an ICC had finally been generated in the international community. The problems faced by the two tribunals in recruiting first-rate judges and prosecutors, providing adequate funding, and gaining custody over war crimes suspects demonstrated the need for a permanent ICC (Sadat 2000). The ILC met again in June 1993, one month after the Security Council had created the ICTY (Von Hebel 1999). The Working Group then wrote a series of draft articles, which the ILC presented to the General Assembly and national governments for written comments. On December 9, 1993, the General Assembly expressed its support for the activities of the ILC through Resolution 48/31, requesting that the ILC prioritize the completion of a draft statute in 1994.

The ILC evaluated two draft statutes in 1994, before deciding on a sixty-article statute (Sadat 2000). Concerned with the political pressures from states, the ILC avoided contentious issues, such as the definition of crimes and the funding of the ICC. On certain issues, including jurisdictional regimes and the ICC's organizational structure, the ILC decided to place primacy on the principle of state sovereignty. The basic premise of the draft statute was that the ILC should complement the proceedings in the national courts, rather than replace them. Moreover, the ICC would only prosecute the most serious cases of international criminal law violations, in situations where national trials would either be ineffective or not be held at all. After some debate within the ILC, it was decided that the ICC would not be responsible for unifying or creating international law, hence, the ICC would not be given any advisory jurisdiction. The draft statute specified that the court would have jurisdiction with regards to both treaty crimes and violations of international humanitarian law. The ICC would only hear cases that were submitted to it by either states parties to the treaty or the UN Security Council. Leila Sadat remarked that "the proposed State consent regime and system of jurisdictional reservations probably would have completely crippled the proposed Court, except in cases involving affirmative action by the Security Council" (Sadat 2000, 39).

The draft statute also described the structure of the ICC. The court would be composed of four organs: a Judiciary with a pretrial and an appellate division, a Registry, a Procuracy, and a Presidency. The ILC envisioned that, with the exception of the Registry, the ICC's organs would function on a periodic basis. Rather than the permanent court lobbied for by human rights activists and certain states, the ILC's draft statute proposed merely a standby court (Sadat 2000).

Upon presenting the draft statute to the General Assembly, the ILC recommended that the Assembly organize an international conference of plenipotentiaries to study the statute and produce a convention on the establishment of an ICC (Von Hebel 1999). Despite the continued reservations of some states on the necessity of an ICC, most of the participants in the Sixth Committee expressed their approval of the draft statute's objective of creating an ICC while simultaneously respecting the principle of state sovereignty. The majority of states concurred that more preparatory work was needed before a diplomatic conference on an ICC could be held. On December 9, 1994, General Assembly Resolution 49/53 set up an Ad Hoc Committee on the Establishment of an International Criminal Court. The Ad Hoc Committee was given the mandate to review the substantive and administrative issues that arose from the ILC's draft statute and plan for an international conference of plenipotentiaries.

Around sixty delegations participated in two meetings of the Ad Hoc Committee in 1995. The Committee became divided over the necessary steps to take. On the one hand, some delegations argued that more general discussions were still needed before a decision on a diplomatic conference could be made, while on the other hand, the like-minded states believed that quick progress could be made by setting up a Preparatory Committee that would prepare a new draft statute, and by holding a conference of plenipotentiaries as soon as 1997. A compromise was then reached within the Ad Hoc Committee, where the members would engage in further debate while simultaneously drafting an ICC convention that would be reviewed by a conference of plenipotentiaries at a later date.

The General Assembly followed up on the report of the Ad Hoc Committee by setting up a Preparatory Committee that was responsible for preparing draft texts (Sadat 2000; Von Hebel 1999). But the General Assembly also postponed any decision on the date and organization of a future conference of plenipotentiaries. The Preparatory Committee was open to all UN member states as well as members of specialized agencies, and was mandated to formulate a widely acceptable draft of an ICC convention for future consideration by a conference of plenipotentiaries. The Preparatory Committee held three, two-week sessions in 1996, during which it collected the various draft proposals. But, according to Herman Von Hebel, "in terms of taking stock of all problems involved, the PrepCom in 1996 could be considered rather productive; in terms of substantive negotiations, that PrepCom only provided a modest first step" (Von Hebel 1999, 34). The divide between the committee participants became evident once again during the final session of 1996, as the like-minded group of states pressed for the holding of a diplomatic conference as soon as 1997, while other countries either proposed 1998 as the earliest possible date for the conference or argued that the time was not yet right for setting a date. Negotiations produced a compromise between the parties, where it was agreed that the Preparatory Committee would convene three or four times for a total of nine weeks during 1997 and the spring of 1998, and that a diplomatic conference of plenipotentiaries would be held later in 1998.

The General Assembly passed Resolution 51/207 on December 17, 1996, declaring that the diplomatic conference would take place in Italy sometime during the summer of 1998 (Von Hebel 1999). Beginning in 1997, the Preparatory Committee decided to no longer record the proceedings of its meetings, in the belief that privacy would facilitate the negotiation of deals between committee members. The sole documents that were produced during the three meetings were the draft articles. On December 15, 1997, General Assembly Resolution 52/160 announced that the conference of plenipotentiaries would be held in Rome from June 15 till July 17, 1998. The Chairman of the Preparatory Committee, Adriaan Bos, who was the Legal Adviser of the Ministry of Foreign Affairs of the Netherlands, organized a January 1998 intersessional meeting in the Dutch city of Zutphen. The meeting included the Bureau of the Committee of the Whole, the chairs of the working groups, the coordinators, and the UN Secretariat. The objective of the meeting was to deal with technical issues, such as the structure and placement of the articles, the amount of detail in the text, inconsistent points, and overlapping material. By the end of the meeting, the draft statute had been completely reworked. The Preparatory Committee then decided to abandon the ILC's draft statute and refine its own version. On April 3, 1998, during its final session, the Preparatory Committee adopted a draft statute that would be presented at the Rome Conference in June.

The Rome Conference

The United Nations Diplomatic Conference of Plenipotentiaries on the Establishment of an International Criminal Court was held in Rome, at the headquarters of the Food and Agricultural Organization, from June 15 till July 17, 1998. One hundred and sixty states, thirty-three intergovernmental organizations, and a coalition of two hundred and thirty-six NGOs participated. The various parts of the Preparatory Committee's draft statute were divided among the different working groups of the Committee of the Whole, the latter of which was given the responsibility of negotiating the statute in its entirety. The negotiations were directed by two competent and effective chairpersons from middle power countries. Adriaan Bos of the Netherlands chaired the Ad Hoc and Preparatory Committees, but fell ill a few weeks before the Rome Conference was to begin. Bos was replaced by Philippe Kirsch, the Legal Adviser of the Canadian Department of Foreign Affairs and International Trade, who assumed the chairmanship of the Committee of the Whole. Mahnoush Arsanjani described how the ICC initiative was facilitated by the leadership of these two men:

> Bos's style, incorporating the most detailed understanding of the positions of various governments and the political dynamics behind them, was reassuring and deliberate, a technique that was useful in keeping all sides engaged during the early phases of the negotiations. Kirsch is a consummate international parliamentarian, and his style is swift and creative in the formation of consensus. He was animated by a determination to assemble a final package by maintaining a consistent focus and negotiating both bilaterally and multilaterally. This style proved to be crucial in forging compromise texts for the statute. (Arsanjani 1999, 24)

The Rome Statute

The draft statute that was presented by the Preparatory Committee consisted of a preamble and thirteen parts, including one hundred and twenty-eight articles. The structure of the statute had been set at the January 1998 Zutphen meeting; the Rome Conference did not address the statute's structure whatsoever. Three principles provided the foundation for the statute (Arsanjani 1999). First, under the principle of complementarity, the ICC may assume jurisdiction only when national legal systems are either unable or unwilling to exercise jurisdiction. In cases where jurisdiction is shared between the ICC and national courts, the latter have primary jurisdiction. The ICC will only act when national courts do not. Second, the statute limits the ICC's jurisdiction to the most serious crimes of concern to the international community. By restricting the caseload of the ICC, it was hoped that the court would not become overburdened by cases that national courts could handle, that the costs of the ICC for the international community would be reduced, and that the court would gain in credibility, effectiveness, and moral authority by earning the acceptance of states. Third, the statute is rooted in customary international law, in order to make it more widely acceptable. While this approach was applied mainly to the definition of crimes, the statute's provisions dealing with the general principles of criminal law and the rules of procedure drew on both common and civil law.

The most important parts of the statute, in terms of substantive humanitarian law, are Articles 6 to 8 dealing with genocide, crimes against humanity, and war crimes (Meron 1999). Article 5(d), covering the crime of aggression, was included in the statute due to a compromise between those parties who insisted that aggression should be treated the same way as the other crimes, and others who stressed that aggression should be excluded since it had not been adequately defined, and because it is a crime committed more frequently by states than individuals. The inclusion of the crime of aggression in the ICC's jurisdiction was made tentative on both the adoption of a definition of the crime, in accordance with Articles 121 and 123, as well as the establishment of conditions under which the ICC would exercise such jurisdiction.

With regards to the crime of genocide, Article 6 is a restatement of Article II of the Convention on the Prevention and Punishment of the Crime of Genocide, adopted by the UN General Assembly on December 9, 1948 (Meron 1999). Acts of genocide include killing or seriously harming members of a national, ethnic, racial, or religious group; deliberately inflicting harsh life conditions on the group with the intent of destroying it; imposing measures that will prevent births within the group; and forcibly transferring the children of the group to another group (*The Rome Statute of the International Criminal Court* 2002 [hereafter *The Rome Statute*]). Furthermore, under Article 25(3)(e) of the Rome Statute, a person may be tried by the ICC for directly and publicly inciting others to commit genocide.

Article 7 of the Rome Statute provides "the first comprehensive multilateral treaty definition of crimes against humanity" (Meron 1999, 49). Eleven categories of crimes against humanity are listed: murder, extermination, enslavement, deportation or forcible transfer of a population, imprisonment in violation of fundamental rules of international law, torture, acts of sexual violence, persecution against groups on the basis of ascribed characteristics, enforced disappearance of persons, the crime of apartheid, and other inhumane acts that intentionally cause great suffering or serious injury (*The Rome Statute* 2002). The inclusion of crimes against women in the Rome Statute redresses a major void in international humanitarian law (Meron 1999).

Article 8 of the statute covers war crimes. The ICC is given jurisdiction when war crimes are committed as part of a policy or plan, or as part of a large-scale commission of these crimes. The Rome Statute provides a long list of acts that may be classified as war crimes. To begin with, grave breaches of the 1949 Geneva Conventions are considered to be war crimes, such as willful killing, torture, the intentional causing of suffering or injury, extensive destruction and appropriation of property, unlawful deportation, transfer, or confinement, compelling a prisoner of war to serve in the hostile military forces, willfully depriving a prisoner of war of the right to a fair trial, and the taking of hostages (*The Rome Statute* 2002). Other examples of war crimes include intentional attacks on civilians and civilian objects, willful attacks on humanitarian personnel, the direct or indirect transfer by an occupying power of parts of its own civilian population onto the territory it occupies, the killing or wounding of individuals in a treacherous manner, the use of poison or poisoned weapons, the utilization of weapons, materials, and methods which may cause superfluous injury or unnecessary suffering, the starvation of citizens as a method of warfare, and the conscription or enlistment of children under fifteen years of age into the military.

Paragraph 2(c) of Article 8 also includes a shorter list of war crimes for armed conflicts not of an international character. These include situations of protracted conflict between governmental authorities and organized armed groups, or solely between the latter. According to Theodor Meron, "the recognition that war crimes under customary law are pertinent to noninternational armed conflicts represents a significant advance" (Meron 1999, 53). Paragraph 2(c) does not apply, however, to situations of internal disturbances and tensions, such as riots and sporadic acts of violence. Meron indicated that there are some shortcomings with the definition of war crimes. Specific references to bacteriological and biological agents and toxins, as well as to chemical weapons, have been deleted from the Rome Statute. Furthermore, the statute does not criminalize the use of any particular weapon in noninternational armed conflicts. Finally, the reference to "protracted armed conflict" in the statute's definition of "non-international armed conflict" implies that the ICC cannot prosecute war crimes that have been committed in situations that fall short of this high benchmark, such as repressive military crackdowns on outbreaks of antigovernment street violence.

The majority of the states which participated in the Rome Conference wanted the ICC to have automatic jurisdiction with regards to genocide, crimes against humanity, war crimes, and the crime of aggression (Arsanjani 1999). But some countries, including the United States, wanted to establish automatic jurisdiction solely for the crime of genocide. They proposed that a consent regime be instituted for the other crimes, where states would agree to either opt in or opt out, or would decide to give their consent on individual cases. In the end, it was decided that the ICC may exercise jurisdiction concerning the crimes enumerated in the Rome Statute, provided that it obtains either the consent of the state on whose territory the crime was committed, or the consent of the state of which the accused is a national. But in situations where a case is referred to the ICC by the Security Council, the ICC will have jurisdiction, even if the crime was committed in a state that is a nonsignatory of the Rome Statute, or the accused is a national of a nonsignatory state. Moreover, in these scenarios, the ICC will exercise jurisdiction even without the consent of the state where the crime was committed or the consent of the state where the accused is a national.

The city of The Hague in the Netherlands was selected as the home of the court. The ICC differs from the International Court of Justice (ICJ) in that the former was designed to try individuals while the latter can only decide disputes between states (Brown 2000). At first, the ICC will have eighteen full-time judges divided into separate chambers which handle trials, pretrial matters, and appeals. The number and status of the judges may be adjusted later depending on the caseload. The judges will be elected by the Assembly of States Parties to a single nine-year term. The selection of judges will take into account equitable geographical representation, representation of the world's major legal systems, a fair representation based on gender, and a consideration of the need for expertise on specific issues. The judges must be nationals of states parties to the treaty (although no two judges may be nationals of the same state), and fluent in either English or French.

The Rome Statute specified that the ICC would come into existence once sixty states had signed and ratified the treaty (Brown 2000). States parties are obligated to cooperate with the ICC to facilitate the investigation and prosecution of crimes within its jurisdiction. Cooperation includes the arrest and transfer of suspects to the ICC, and the provision of evidence to the court. Since states parties must ensure that national laws allow for cooperation with the ICC, most states need to adjust their domestic legislation before ratifying the statute. If a state party denies a request from the ICC for cooperation, it must furnish the ICC with reasons for the denial. A state party may refuse to provide the ICC with evidence that it considers adverse to national security interests, and it may even refuse to provide reasons for a denial of cooperation if the reasons would also threaten national security. If the ICC views a state party's refusal to cooperate as a violation of the statute, it may refer the matter to the Assembly of States Parties, or to the Security Council if the case is based on a referral from that body. But the Rome Statute does not specify what measures may be taken to punish noncompliance.

ICC investigations may be initiated by a referral from either a state party or the Security Council to the ICC Prosecutor, or by the Prosecutor on their own authority (*proprio motu*) based on information that crimes have been committed (Brown 2000). The Prosecutor may obtain additional information from states, international organizations, and NGOs. If an investigation is initiated by the ICC Prosecutor, it must be authorized subsequently by a majority vote of the three judges in the Pre-Trial Chamber. Any orders or warrants requested by the Prosecutor for the purpose of the investigation must also be authorized by these judges. Victims are permitted to submit their views and information to the Pre-Trial Chamber. The Security Council may suspend any ICC investigation or prosecution for a renewable period of twelve months by adopting a resolution under Chapter VII of the UN Charter.

Under the Rome Statute, a three-judge Trial Chamber presides over the trial, which is held in the presence of the accused. International human rights law provides the accused with certain rights, including "the presumption of innocence, the right to a public hearing, the right to counsel, the right to a speedy trial, and the right to compel the attendance of witnesses on the same terms as the prosecution" (Brown 2000, 76–7). The statute does make some new contributions to international criminal law, such as the provision of special measures to ensure the protection of victims and witnesses whose testimony before the ICC may endanger them, and the ability of the ICC to order that reparations be made to the victims of a person who is convicted. During the trial, the Prosecutor is responsible for proving that the accused is guilty beyond a reasonable doubt. The decisions of the Trial Chamber are made by a majority of the three judges who preside. According to the ICC appeals process, either a conviction or an acquittal may be appealed to the five-judge Appeals Chamber. The latter body may either reverse or amend the verdicts of the Trial Chamber, or it can order that a new trial should be conducted before a new Trial Chamber. The convicted may be sentenced to imprisonment for up to thirty years or to a term of life imprisonment. A convicted criminal would be imprisoned in a country chosen by the ICC from a list of states which have agreed to accept prisoners. The ICC may impose a fine or a forfeiture of assets which were obtained from a crime, and may order that such assets be transferred to a trust fund which benefits the victims of the crime and their families.

The primary sources of funding for the ICC are contributions from states parties, which are based on the scale of assessments used for the UN regular budget (Brown 2000). The United States made the case during the Rome Conference that nonparties to the treaty should not be forced to fund the ICC through their contributions to the UN. The Rome Statute addressed this point by stating that the ICC may receive UN funds to cover its expenses in situations where cases are referred to the ICC by the Security Council. The ICC may also accept donations from national governments, international organizations, corporations, individuals, and other sources. The criteria for voluntary funding are to be decided by the Assembly of States Parties.

Fast-track diplomacy and the ICC initiative

Nongovernmental organizations campaign for an ICC

In February 1995, the NGO Coalition for an International Criminal Court (CICC) was established, in order to coordinate NGO action on the ICC initiative and to disseminate information on the progress of the negotiations (Berg 1997; Pace 1999; Pace and Schense 2001). The CICC consisted of twenty-five organizations at first, but expanded to include around eight hundred NGOs worldwide by the start of the Rome Conference and over a thousand NGOs by June 2000. Around 450 representatives from 235 NGOs were accredited by the UN General Assembly to participate in the Rome Conference. According to the Convener of the CICC, William Pace, nearly all of these NGOs were members of the CICC (Pace 1999). William Pace and Jennifer Schense (2001) indicated that approximately five hundred NGOs may have been represented at the Rome Conference, if one were to count the individual members of NGO coalitions, such as the World

Federalist Movement (WFM) and the Women's Caucus for Gender Justice. In fact, the CICC was the largest delegation overall at the Rome Conference. The WFM delegation, which served as the Secretariat of the CICC, exceeded even the largest delegations from national governments (Pace 1999). Most of the NGOs, however, sent merely one or two representatives, who were able to attend only a portion of the five-week conference.

An informal Steering Committee coordinates the activities of the CICC.[3] The engine of the coalition is its vast web of national and regional networks.[4] The NGOs which make up the CICC are located all over the world, and are concerned with diverse issues, including the environment, the rights of women and children, indigenous peoples, religion, ethics, peace, disarmament, and humanitarian and international law (Berg 1997; Pace and Schense 2001). This broad coalition of NGOs was united under a mandate to cooperate in order to support the establishment of an effective and just ICC. Accolades for the successful efforts of the CICC have come from national governments, UN Secretary-General Kofi Annan, and media experts.

The CICC engaged in numerous activities during the Rome Conference (Pace 1999). The coalition created thirteen working groups which reviewed the 128 articles of the statute, and assisted NGO experts from less developed countries to attend the conference. The CICC organized regional caucuses, including the tricontinental alliance established by groups from Africa, Asia, and Latin America, and convened sectoral caucuses, which dealt with the link between justice and issues like gender, children, and religion. Reports and documents were written and translated by the CICC for the use of NGOs and national governments. The coalition also organized three separate news teams to furnish the conference participants with two daily newspapers and an online bulletin. The CICC provided experts and interns to assist government delegations, and helped coordinate between the conference proceedings and national NGO networks. Furthermore, the CICC briefed the international and regional press on a regular basis, conducted daily strategy sessions, held weekly meetings with the Chair of the Rome Conference, Philippe Kirsch from Canada, and convened regular meetings with national governments, particularly the sixty members of the Like-Minded Group of Countries (LMG). In addition, the coalition produced statistical analyses of the degree of support of government delegations for particular elements of the ICC, which assisted the Rome Statute negotiations considerably.

According to William Pace, "the highly publicized and praised contributions of NGOs at the Rome Conference represented a small fraction of the work done by the Coalition during the previous three and a half years of preparatory meetings" (Pace 1999, 204). Prior to Rome, the CICC conducted an information campaign, met with national governments, prepared background reports, assisted with translations, provided the media with briefings, created regional and sectoral working and support groups, and convened intersessional meetings. The NGOs Amnesty International, Human Rights Watch, Lawyers Committee for Human Rights, and the Women's

Caucus for Gender Justice were particularly adept in preparing documentation and campaign materials for every preparatory meeting leading up to the Rome Conference. The CICC also benefited from the astute leadership of its Convener, William Pace. Two other figures deserve a special mention. Professor Cherif Bassiouni headed the International Superior Institute of Criminal Science in Italy, which helped convene unofficial intersessional meetings during the preparatory period. The Italian politician Emma Bonino led the Transnational Radical Party and the NGO No Peace Without Justice, two groups whose international campaign helped secure a commitment from Italy to host the ICC conference, and which organized a series of meetings with politicians and governments around the world in order to drum up support for the ICC.

The Like-Minded Group of Countries and the ICC

In 1994, the LMG was formed by around a dozen states, who wished to campaign for the convening of a diplomatic conference of plenipotentiaries in 1998. The LMG is an informal association without a permanent membership. By the time of the Rome Conference, approximately sixty countries, most of them middle powers and small states, had joined the group. By June 2000, sixty-seven states were members.[5]

During the meetings of the Preparatory Committee prior to the 1998 Rome Conference, key members of the CICC believed that the outcome of the negotiations on an ICC would depend on the leadership and negotiating capabilities of the LMG (Pace 1999). The LMG managed to assume significant leadership positions at the Preparatory Committee meetings, with the aid of the chairmen of the committee, who represented middle power governments. The first Chairman of the Preparatory Committee, Adriaan Bos from the Netherlands, appointed mainly leaders from the LMG as "issue coordinators." Bos's replacement, Philippe Kirsch from Canada, would continue this strategy of consolidating the influence of the like-minded states in the negotiations. The LMG would also control the proceedings of the Rome Conference by chairing most of the working groups and assuming membership in the Bureau of Coordinators, the executive body that administered the daily affairs of the conference (Schabas 2011). In addition, Philippe Kirsch was elected president of the Rome Conference's Committee of the Whole.

In 1997, the CICC requested that the LMG identify guiding principles that would serve as the foundation for the pro-ICC bloc during the negotiations. At the penultimate session of the Preparatory Committee in December 1997, the LMG reached a consensus on six main principles: the ICC should not be subjected to the oversight of the UN Security Council; the ICC Prosecutor should be independent; the ICC jurisdiction should be extended to cover the crime of genocide, crimes against humanity, war crimes, and the crime of aggression; states should cooperate fully with the ICC; the ICC should make the final decision on issues of admissibility; and a diplomatic conference of plenipotentiaries should be convened in Rome (Pace 1999).

Scholars agree that the alliance between the CICC and the LMG demonstrated the efficacy of soft power and the "new diplomacy" (Pace 1999; Robinson 2001). With the Canadian government at the helm, the members of the LMG were urged to coordinate their positions on both issues of substance and strategy.[6] The LMG had to remain cohesive in order to overcome the antagonism of certain countries—including China, France, India, Mexico, the United Kingdom (prior to the emergence of the Labour government in 1997), and the United States—who were either opposed to or indecisive about the ICC initiative, and refused to set a date for a diplomatic conference.

The Convener of the CICC, William Pace, and other NGO leaders approached the Canadian Minister for Foreign Affairs Lloyd Axworthy for his support on the ICC initiative (Robinson 2001). Axworthy, who had been overwhelmingly successful in achieving the Ottawa Convention banning antipersonnel landmines in 1997, used his bilateral and multilateral contacts, as well as public statements, to spread the word on the necessity of an ICC. The Canadian foreign minister would be an active participant at the Rome Conference, where he lobbied states to remain firm in their will to establish an effective and worthwhile ICC, and contacted other foreign ministers to discuss particular issues at critical stages of the negotiations.

Pressure from the LMG resulted in the Rome Statute's inclusion of war crimes in internal armed conflicts, despite the initial vocal opposition of a few states (Robinson and Oosterveld 2001). Canada campaigned, with success, for the criminalization of sexual and gender-based offenses, including rape, sexual slavery, enforced prostitution, and persecution on the basis of gender. Furthermore, after fierce negotiations, the conference participants finally accepted a Canadian-proposed definition of crimes against humanity, which stated that these crimes are not only punishable when committed during armed conflict, but also when they are committed during incidents of societal disturbance (e.g. riots) or in times of peace.

After the Rome Conference's five weeks of negotiations, there were still stalemates with regards to the scope of the ICC's jurisdiction, the degree of the ICC's independence, the extent to which ratification would automatically provide the court with competence over all the crimes in the statute, and whether the ICC Prosecutor would be allowed to initiate an investigation without a referral from a state party (Robinson 2001). To bridge the divide, the Bureau of Coordinators drafted a final proposal which reflected the LMG's wish for a strong ICC, but also accommodated the concerns of the minority of detractors at the conference. This package deal was widely endorsed by the delegations on the final day of the conference, despite a couple of last minute attempts to sabotage the treaty. First, India proposed that the use of, or threat to use, nuclear weapons should be considered a war crime (Anbarasan 1998; Weschler 2000). But Norway moved to table the motion, which was seconded by Malawi and Chile, and India lost the resulting vote with 114 states against, sixteen for, and twenty abstentions.

Second, the United States objected to the Rome Statute, because it gives the ICC the authority to exercise jurisdiction over American nationals even without the consent of the United States (Leigh 2001). At the final meeting of the Committee of the Whole on July 17, 1998, David Scheffer, the US Ambassador-At-Large for War Crimes Issues and head of the American delegation at the Rome Conference, proposed an amendment to the treaty, which would have made the ICC jurisdiction conditional on both the consent of the state on whose territory the crime was committed, and the consent of the state of which the accused is a national (Weschler 2000, 107). But Norway immediately tabled the motion, which Sweden and Denmark seconded. A vote was then held, and the American proposal was soundly defeated, with 113 states voting against, seventeen for, and twenty-five abstentions ("The Birth of a New World Court" 1998).

The conference participants proceeded to adopt the Rome Statute on July 17, 1998, with 120 states voting in favor, seven against, and twenty-one states abstaining (Robinson 2001). At the request of the United States the vote was not recorded ("Permanent International Criminal Court Established" 1998). There is some disagreement over which seven states voted against the Rome Statute. Although it is widely accepted that the United States, China, Israel, and Libya cast negative votes, David Bosco (1998) claimed that Algeria, Qatar and Yemen also opposed the statute. William Nash (2000) singled out Iraq, Qatar, and Yemen, while Monroe Leigh (2001) identified Iran, Iraq, and Sudan as the three remaining dissenters. Ironically, the United States joined the company of a few pariah states in voting against the ICC.

The Preparatory Committee for the ICC held an additional five meetings over the following two years. States which opposed the Rome Statute tried repeatedly to have the treaty amended. But spurred by a fear that any renegotiations would weaken the Rome Statute, the LMG and the CICC succeeded in preserving the statute in its present form (Pace and Schense 2001). Skilled diplomacy by the pro-ICC bloc helped get the draft *Rules of Procedure and Evidence and the Elements of Crimes* adopted by consensus on June 30, 2000. The CICC launched a promotional campaign to encourage national governments to enact domestic legislation that would implement the ICC. The campaign bore fruit, as the sixtieth ratification of the Rome Statute was deposited with the UN on April 11, 2002, and the ICC came into effect on July 1st of that year ("International Criminal Court Statute Becomes Effective" 2002). As of September 2011, 117 countries had become states parties to the Rome Statute (ICC 2011d). Through their competent leadership on the initiative to establish the ICC, the middle powers addressed the global demand for justice, and managed to fill a void in the realm of human security.

The United States and the ICC

US policy and the ICC negotiations

The United States was an early advocate for the creation of an ICC (Pfaff 1998). Secretary of State Madeleine Albright emphasized that, by prosecuting individuals who carry out atrocities, an ICC would deter war crimes from occurring in the future. The Bill Clinton administration adopted the position that the United States would support the court, but only if the United States were exempt from its jurisdiction. The administration pursued three main objectives during the ICC negotiations (Scheffer 1999). First, it was desired that the negotiations would result in a treaty. Second, the Clinton administration argued that the ICC had to take into account American responsibility for maintaining international peace and security. Third, the administration opposed the establishment of an independent ICC Prosecutor. Ambassador David Scheffer summarized the Clinton administration's position succinctly:

> Since 1995, the question for the Clinton administration has never been whether there should be an international criminal court, but rather what kind of court it should be in order to operate efficiently, effectively and appropriately within a global system that also requires our constant vigilance to protect international peace and security. At the same time, the United States has special responsibilities and special exposure to political controversy over our actions. This factor cannot be taken lightly when issues of international peace and security are at stake. We are called upon to act, sometimes at great risk, far more than any other nation. This is a reality in the international system. (Scheffer 1999, 12)

In early 1993, the Clinton administration launched a review of the draft proposal for an ICC which the ILC had been discussing since 1992. The American suggestions for the proposed ICC included a considerable role for the UN Security Council in referring cases to the ICC, the elaboration of an adequate definition of war crimes in the Rome Statute, the exclusion of drug trafficking and the crime of aggression—which was difficult to define—from the statute, and the further study of US apprehensions about the inclusion of crimes of international terrorism in the statute (Scheffer 1999).

Washington was concerned that the ICC would be used as a forum for prosecuting US military personnel who serve abroad. Critics of the ICC pointed to the statements made by some Russian and Serbian leaders, who had suggested that the United States should be tried for its aerial bombing during the Kosovo intervention ("International Justice" 2001). The United States was also worried that the ICC would not guarantee American military personnel their constitutional rights to a jury trial and due process, which are protected under the Fifth and Sixth Amendments to the US Constitution (Everett 2000). According to the Fifth Amendment,

> No person shall be held to answer for a capital, or otherwise infamous crime, unless on a presentment or indictment of a Grand Jury, except in cases arising in the land or naval forces, or in the Militia, when in actual service in time of War or public danger; nor shall any person be subject for the same offence to be twice put in jeopardy of life or limb; nor shall be compelled in any criminal case to be a witness against himself, nor be deprived of life, liberty, or property, without due process of law; nor shall private property be taken for public use, without just compensation. ("The Constitution of the United States of America" [1787] 2002, 261–2)

The Sixth Amendment states:

> In all criminal prosecutions, the accused shall enjoy the right to a speedy and public trial, by an impartial jury of the State and district wherein the crime shall have been committed, which district shall have been previously ascertained by law, and to be informed of the nature and cause of the accusation; to be confronted with the witnesses against him; to have compulsory process for obtaining witnesses in his favor, and to have the Assistance of Counsel for his defence. ("The Constitution of the United States of America" [1787] 2002, 262)

Determined to prevent any politically motivated trials of American soldiers, as well as to protect their constitutional rights, the United States insisted that the ICC should be subjected to UN Security Council controls over which cases it may pursue (Omestad 1998). As a permanent member of the Security Council, the United States would thereby retain a veto on the activities of the ICC.

It has been argued by some scholars that the ICC does not pose a genuine threat to American constitutional rights. Since the Fifth Amendment of the US Bill of Rights excludes service members from the guarantee of a jury trial in a time of war or public danger, the fact that the Rome Statute does not provide the accused with the right of a trial by jury becomes less of an issue (Leigh 2001; "U.S. Signing" 2001). Furthermore, under the complementarity regime of the ICC, the court may assume jurisdiction only when a national legal system is either unable or unwilling to launch an investigation (Carter 2002; Leigh 2001; Tepperman 2000). This scenario should never occur in the case of the United States and other democracies with effective judicial systems. But Ambassador Scheffer emphasized that the complementarity regime does not offer sufficient protection for American citizens:

> Even if the United States has conducted an investigation, again as a nonparty to the treaty, the court could decide there was no genuine

> investigation by a 2-to-1 vote and then launch its own investigation of U.S. citizens, notwithstanding that the U.S. Government is not obligated to cooperate with the ICC because the United States has not ratified the treaty. (Scheffer 1999, 19)

The United States was therefore committed to ensuring that the ICC would not threaten the constitutional rights of the American citizenry. The final draft treaty produced by the ILC fulfilled the objectives of the United States to some degree (Scheffer 1999). The draft statute acknowledged that cases which concern the Security Council's functions under Chapter VII of the UN Charter should only be addressed by the ICC after a Security Council referral, that the prosecution of crimes of aggression should require the prior approval of the Security Council, and that the ICC Prosecutor should only act when a case is referred by either a state party or the Security Council. Furthermore, the draft statute included a provision allowing a state party to opt out of one or more of the categories of crimes when ratifying the treaty, thereby abolishing the ICC's jurisdiction over the state party's citizens for those particular offenses.

During the meetings of the Preparatory Committee, the US delegation helped draft the trial procedures of the ICC, as well as define the rights of defendants (Roth 1998). The Rome Statute includes numerous provisions which guarantee due process, such as the right of confrontation and cross-examination, the right to remain silent, the presumption of innocence, the right to assistance of counsel, protection against double jeopardy, privilege against self-incrimination, the right to be present at trial, the prohibition of trials in absentia, and the exclusion of evidence that was obtained illegally. The inclusion of these protections for the accused may be credited in part to tough negotiating by the US delegation during the meetings of the Preparatory Committee. In the words of Ambassador Scheffer:

> Due process protections occupied an enormous amount of the U.S. delegation's efforts. We had to satisfy ourselves that U.S. constitutional requirements would be met with respect to the rights of defendants before the court. Parts 5–8 of the treaty contain provisions advocated by the U.S. delegation to preserve the rights of the defendant and establish the limits of the prosecutor's authority. (Scheffer 1999, 17)

The American delegation also succeeded in getting the procedures of the ICC restructured, and pushed for the broadening of the complementarity regime, to include a deferral to national jurisdiction as soon as a case has been referred to the ICC (Scheffer 1999). The United States worked with the other permanent members (P5) of the UN Security Council—China, France, Russia, and the United Kingdom—to derive an acceptable definition of the crime of aggression, where only a person who could direct or control the political and military actions of a state may be investigated for such a crime.

The American delegation also joined the LMG in revising the definition of crimes against humanity, to include crimes committed during intrastate wars and in the absence of armed conflict.

The United States insisted at the Rome Conference that the Security Council's authorization should be required for each ICC prosecution (Brown 2000). This would have enabled each of the P5 states to veto any ICC prosecution they disagreed with. But to the dismay of the American delegation, the LMG used its influence in the Bureau of Coordinators—bolstered by the appointment of LMG officials as issue coordinators—to secure the establishment of an independent ICC Prosecutor, who is authorized to initiate investigations and prosecutions of crimes without requiring a referral from a state party or the Security Council (Scheffer 1999). Thus, the United States was unsuccessful both in preventing the ICC Prosecutor from being empowered to launch investigations independently, and in ensuring an authoritative position for the Security Council with regards to the selection of the criminal cases that the ICC would investigate.

In addition, the United States was upset with the LMG's control over the negotiations at the Rome Conference. Ambassador Scheffer complained that "the process launched in the final forty-eight hours of the Rome Conference minimized the chances that [the] proposals and amendments to the text that the US delegation had submitted in good faith could be seriously considered by delegations" (Scheffer 1999, 20). During this period, the draft statute was revised, behind closed doors, by a small number of delegates, most of whom were from the LMG. These delegates brokered deals with holdout governments in order to convince them to support a draft that was finalized at 2:00 a.m. on July 17th, the last day of the conference. The LMG's "take it or leave it" approach involved the rewriting of significant portions of the statute, without subjecting the text to review by either the Drafting Committee or the Committee of the Whole. Ambassador Scheffer acknowledged that the US delegation was at a disadvantage compared to the LMG at the Rome Conference, because "the United States usually had to build support for its positions through time-consuming bilateral diplomacy" (Scheffer 1999, 15).

The final draft of the Rome Statute included a provision, unacceptable to the United States, whereby if the treaty were amended in the future to include a new crime or a redefinition of an existing crime, then states parties would be permitted to immunize their nationals from prosecution for this new crime, but the nationals of states not parties would remain subject to potential prosecution (Scheffer 1999). Moreover, to the chagrin of the United States the Rome Statute included the crime of aggression, despite the fact that no consensus on a definition of this crime had been reached at the conference.

On the final day of the conference, the US delegation made a couple of proposals that would have weakened the Rome Statute (Arsanjani 1999). First, Ambassador Scheffer argued that the ICC's jurisdiction should require the consent of both the state where the crime was committed and the state of nationality of the accused. The United States was concerned that American military personnel serving overseas would be prosecuted by the ICC, even if the United States was not a party to the Rome Statute. The American delegation claimed that an overextension of the ICC's jurisdiction would force the United States to reconsider the deployment of its military abroad, including the performance of alliance obligations and humanitarian interventions. But the majority of the conference participants believed that the requirement of double consent would paralyze the ICC and rejected this proposal.

Second, the United States suggested that the ICC should not have jurisdiction in cases where the state of nationality of the accused declares that the crime was committed while the accused was fulfilling an official duty. If so, responsibility for the criminal act would shift from the individual to the state, where general international law, rather than the Statute of Rome, would be applicable. But the other conference participants rejected this proposal as well prior to the vote on the Rome Statute.

After the Rome Conference

On July 23, 1998, less than a week after the conclusion of the Rome Conference, Ambassador Scheffer presented the US Senate Committee on Foreign Relations with a summary of the American delegation's objections to the Rome Statute (Frye 1999). First, the US delegation disagreed with the parameters of the ICC's jurisdiction. Under Article 12 of the statute, the ICC has jurisdiction when either a crime is committed on the territory of a state party, or when the accused is a national of a state party. Citizens of states not parties to the Rome Statute, such as the United States could still be prosecuted by the ICC. In a possible scenario, if a state not party participated in a peacekeeping mission on a state party's territory, the peacekeeping personnel could fall under the ICC's jurisdiction. Ambassador Scheffer made the additional argument that states not parties may actually be more vulnerable to the ICC war crimes jurisdiction than states parties, since the former cannot opt out of the war crimes jurisdiction for a seven-year period like states parties are permitted.

Second, the United States and the other members of the P5 of the Security Council were unable to convince the rest of the conference participants to adopt their proposal allowing states to opt out of the ICC jurisdiction on both crimes against humanity and war crimes for up to ten years (Frye 1999). The American delegation felt that such a lengthy opt-out period would give states time to evaluate if the ICC was operating in an effective and impartial manner. Article 124 of the Rome Statute allows a seven-year opt-out period, but only on the issue of war crimes. The P5 proposal would have also protected nonsignatory states from the ICC's jurisdiction, unless the Security Council decides otherwise in a particular case. Unfortunately for the P5, their proposal was rejected by the LMG, the dominant bloc at the Rome Conference (Scheffer 1999).

Third, the United States objected to the establishment of an independent ICC Prosecutor. Fourth, the American delegation was dissatisfied that the Rome Conference participants included the crime of aggression in the statute, but set aside the definition of the crime for future negotiations. The US preferred that the Security Council determine when an incident of aggression has occurred, and if the case should be referred to the ICC. Finally, the US delegation was disappointed with the adoption of a provision which prohibits reservations to the Rome Statute.

Despite its opposition to the Rome Statute, Washington still sought a means by which it could influence the future development of the ICC. On December 31, 2000, just hours before the deadline, the US reversed course and signed the Rome Statute ("Sign On, Opt Out" 2001). President Clinton explained that by signing the treaty, the United States was demonstrating its moral leadership on the ICC issue. But there was a greater calculation behind the American signature: the United States would receive an invitation to participate in the subsequent technical meetings that would be convened to work out the Rome Statute's details. The American delegation would then be able to push for the adoption of provisions which correspond with US interests. Chief among these clauses would be an exemption for states which do not ratify the statute. President Clinton emphasized that he would not ask the US Senate to ratify the Rome Statute in its present form. Moreover, Clinton knew that US ratification would have been impossible, as the Chairman of the Senate Armed Services Committee, Senator Jesse Helms (R-North Carolina), had declared that any treaty for an ICC that could prosecute US citizens would be "dead on arrival" in the Senate (Cassel 2001, 14).

Senator Helms, Representative Henry Hyde (R-Illinois), and House Majority Whip Tom DeLay (R-Texas) introduced the American Service-Members' Protection Act (ASPA) of 2000 in Congress, as an amendment to the Department of Defense Appropriations Act (Eviatar 2001; "U.S. Signing of the Statute of the International Criminal Court" 2001 [hereafter "U.S. Signing"]). The ASPA was intended to prohibit any U.S. court or government from cooperating with the ICC, prevent American forces from participating in UN-sponsored military operations unless they are granted immunity from criminal prosecution by the ICC, and cut off US aid to all states parties to the Rome Statute, with the exception of NATO members, key non-NATO allies, and states which agree not to transfer US personnel to the ICC. Under the legislation, the president would be authorized to use whatever means necessary to free American personnel from ICC captivity, which inspired some to refer to the ASPA as the "Hague Invasion Act" (Roach 2006). While the ASPA did not receive sufficient votes in the US Senate in September 2000, it was subsequently adopted by the Senate in a second vote on December 7, 2001, and became part of the 2002

Supplemental Appropriations Act for Further Recovery From and Response to Terrorist Attacks on the United States.

The ASPA received considerable criticism for its intent to subvert multilateral cooperation by promoting unilateralism in US foreign policy. Doug Cassel indicated that the Republicans had a strong motivation for the legislation: "if [the ASPA] seems calculated to offend other nations and to thwart U.S. participation in international peacekeeping, so much the better" (Cassel 2001, 14). Daniel Benjamin remarked that, "in the ICC project, American right-wingers hear the whir of black helicopters, the approach of world government and the loss of U.S. sovereignty" (Benjamin 2001, 31).

The George W. Bush administration adopted a belligerent position vis-à-vis the ICC. On September 25, 2001, the Bush administration announced its support for the ASPA, after a provision was included which permitted the president to provide military assistance to a state party to the Rome Statute if he considers it to be in the national interest. In May 2002, Bush "unsigned" the Rome Statute, the first time ever a US president revoked the signature of a former president on a treaty (Anderson 2002; Meyer 2002). That same month, the Undersecretary of State for Arms Control and International Security, John R. Bolton, sent a letter to Kofi Annan indicating that the United States would not ratify the Rome Statute and therefore had no legal obligations stemming from President Clinton's signature in December 2000 (Murphy 2010).

The United States surprised many in the international community when it threatened to veto the routine renewal of all UN peacekeeping operations, starting with the mission in Bosnia, if the Security Council did not grant a permanent immunity from ICC prosecution to all UN peacekeepers ("Both Sides Lose" 2002). But each of the other Security Council members, including both ICC supporters and states like China and Russia which have been unwilling to join the court, refused to pass such a resolution. The United States then pressed, unsuccessfully, for a twelve-month exemption from prosecution for UN peacekeepers that would be automatically and perpetually renewable. The United Kingdom finally brokered the compromise Security Council Resolution 1422, where immunity was extended, for a period of twelve months that would be renewable annually, to all participants in either UN or UN-authorized operations who are from countries which are not parties to the Rome Statute. The United States had to settle for an exemption that was not permanent, and could have been vetoed by any of the P5 when it came time for its annual renewal.[7]

The Bush administration withdrew from all negotiations to establish the ICC. Furthermore, on August 3, 2002, President Bush signed the ASPA into law (HRW 2002). By May 2005, the US government had negotiated Bilateral Immunity Agreements (BIAs) with one hundred countries, whereby the parties agreed not to transfer each others' citizens to the custody of the ICC (American Non-Governmental Organizations Coalition for the International Criminal Court 2002a, 2004 [hereafter AMICC]). At least thirty-five states, which failed to sign BIAs, had their US military aid suspended (Lobe 2003a, 2003b). But after the ASPA came into effect, President Bush waived its prohibition on US military assistance for several ICC states parties which had not signed BIAs, in cases where it was expedient for American political interests to do so. On November 14, 2005, President Bush signed the controversial Nethercutt Amendment to the Foreign Operations, Export Financing, and Related Programs Appropriations Act (AMICC 2002b, 2004). The amendment, introduced by Representative George Nethercutt (R-Washington), suspended Economic Support Fund aid to ICC states parties which had neither signed BIAs with the United States nor received a presidential waiver.[8] But despite American antagonism, the ICC became a reality in July 2002, and has begun investigations and trials of criminal cases.

There has been a considerable softening of the US position on the ICC in recent years, particularly since the ICC began proceedings regarding the situation in Darfur, Sudan (ASIL 2009). The United States abstained on the 2005 Security Council resolution which referred the situation in Darfur to the ICC Prosecutor. In October 2006, President Bush waived the US prohibition on military assistance to twenty-one ICC states parties, mainly because the prohibition hindered international cooperation on the "War on Terror" and the "War on Drugs" (ASIL 2007). The prohibition on military aid to ICC states parties was totally lifted in 2008 (Murphy 2010).

Although the Barack Obama administration has decided not to sign the Rome Statute, it has adopted a strategy of "positive engagement" with the ICC. John Murphy (2010) suggested that the change in the US position was due to a positive assessment of the ICC's professionalism in dealing with more than 240 complaints about the US-led operations in Iraq. The United States was satisfied with Prosecutor Luis Moreno-Ocampo's decision not to open an investigation, due to a lack of ICC jurisdiction (in the case of American soldiers serving in Iraq, as both the United States and Iraq are nonparties to the Rome Statute) and a lesser gravity of war crimes (in the case of the dozen or fewer victims of British war crimes in Iraq, as the United Kingdom is a state party).

The Obama administration sent a delegation to observe the ICC Assembly of States Parties in November 2009, and expressed its encouragement of the ICC's investigation in Darfur, but also its concern with proposals to define the crime of aggression at the 2010 Review Conference (Crook 2010). An American observer delegation was also sent to the Kampala Review Conference, held from May till June 2010 (ASIL 2010). The states parties reached a consensus on a definition of aggression as "a crime committed by a political or military leader which, by its character, gravity and scale constituted a manifest violation of the [UN] Charter" (ASIL 2010, 512). An act of aggression may be referred to the ICC by the Security Council, or the ICC Prosecutor may launch an investigation upon request from a state party or on their own initiative, with prior authorization from a PTC. The ICC will not have jurisdiction if the crime of aggression was committed on the territory of a nonstate party or by their nationals, or if a state party declares that it does not accept ICC jurisdiction with regards to aggression. Furthermore, the ICC will not have jurisdiction on aggression until states parties have adopted it through consensus or a two-thirds majority sometime after January 1, 2017. The United States was highly satisfied with this outcome of the Kampala conference.

The ICC at work

The first conference of the states parties to the Rome Statute was held in September 2002, and the inaugural session of the ICC, featuring the swearing in of the eighteen judges, was conducted on March 11, 2003 (Levene 2003). The Canadian Philippe Kirsch, who had served astutely as the chair of the Rome Conference, was appointed as the ICC's first President for a term of six years. He was replaced in March 2009 by the South Korean judge Sang-Hyun Song, who is currently serving a three-year renewable term as ICC President. Luis Moreno-Ocampo from Argentina was elected by the states parties to a nine-year term of ICC Prosecutor, which began on June 16, 2003 (ICC 2011h).

As of September 2011, five situations had been referred to the Office of the Prosecutor, three by states parties to the statute, and two by the UN Security Council. The ICC Prosecutor also opened his own investigation *proprio motu* into the situation in Kenya. In addition, the Office of the Prosecutor was conducting preliminary investigations in Afghanistan, Colombia, Côte d'Ivoire, Georgia, Guinea, Honduras, Nigeria, Palestine, and South Korea (ICC 2011f).

Democratic Republic of the Congo

The situation in the Democratic Republic of the Congo (DRC) became the ICC's first case. Since conflict erupted in the DRC in 1994, millions of civilians have been killed, raped, tortured, and displaced. Prosecutor Moreno-Ocampo announced in July 2003 that he would monitor events in the DRC closely, particularly crimes committed in the Ituri region, which had suffered through an ethnic conflict between the Lendu and Hema tribes since 1999 (ICC 2004c). In September 2003, Moreno-Ocampo declared to the Assembly of States Parties that he was considering submitting a request to a PTC for authorization to begin his own investigation of crimes in the DRC. He emphasized, though, that a referral and active support from the DRC government would facilitate his work. The president of the DRC, Joseph Kabila, responded in November 2003 with a letter to Moreno-Ocampo welcoming the involvement of the ICC. On March 3, 2004, Kabila sent the Prosecutor another letter, which referred to the ICC the situation of crimes committed on the territory of the DRC since July 2002, when the jurisdiction of the court began (ICC 2004d).

Prosecutor Moreno-Ocampo announced on June 23, 2004, that he would launch an investigation of crimes committed in the DRC (ICC 2006a). That same day, ICC President Philippe Kirsch assigned the situation of the DRC to the PTC I. On January 13, 2006, Moreno-Ocampo filed an application for a warrant of arrest, which the PTC I proceeded to issue under seal in February and made public in March. Thomas Lubanga Dyilo, the founder of the *Union des Patriotes Congolais* (Union of Congolese Patriots, UPC) and its military wing, the *Force Patriotique pour la Libération de Congo* (Patriotic Force for the Liberation of Congo, FPLC), was charged with the war crimes of enlisting, conscripting, and using children under the age of fifteen to participate actively in hostilities in the Ituri region of the DRC between September 2002 and August 13, 2003.

Arresting Lubanga Dyilo proved to be relatively easy, due to the fact that he had been in the custody of Congolese authorities since August 2003 (ICC 2006b). Lubanga Dyilo was promptly transferred to the ICC detention center in The Hague on March 17, 2006, and made his first appearance before the PTC I three days later. A confirmation hearing was held from November 9–28, 2006, where the PTC I heard the submissions of the participants, as well as the arguments of the legal representatives of the victims (ICC 2007a). On January 29, 2007, the PTC I announced that it had confirmed the three charges brought against Lubanga Dyilo by the Prosecutor, and was referring the case for trial before a Trial Chamber. The following day, Lubanga Dyilo proceeded to appeal the confirmation of charges against him, but on June 13, 2007, the Appeals Chamber of the ICC dismissed his appeal (ICC 2007d). Lubanga Dyilo's trial began on January 26, 2009 (ICC 2011c). The trials of two additional Congolese rebel leaders began on November 24, 2009, while a fourth suspect was arrested by French authorities in October 2010. An arrest warrant was issued for a fifth rebel leader in 2006, but he still remains at large. On March 14, 2012, Trial Chamber I found Lubanga Dyilo guilty of war crimes, the first ever verdict reached by the ICC (ICC 2012).

Uganda

The second situation that the ICC addressed related to the activities of the insurgent Lord's Resistance Army (LRA) in Northern Uganda, a region that had suffered violent conflict since 1986 (GlobalSecurity.org 2006; ICC 2005c, 2005d). Led by Joseph Kony, the LRA aimed to overthrow the Ugandan government and institute a regime based on the biblical Ten Commandments. The group became a transnational problem in 1994, when they established bases across the border in Southern Sudan, from where they organized guerrilla attacks on communities in both Sudan and Uganda.

The LRA also operated in the Bunia region in the Eastern part of the DRC, allying itself with various rebel groups in the area.

While the LRA claimed that they were fighting for the freedom of the Acholi people of Northern Uganda, most of their attacks were actually on Acholis (ICC 2005c). Over a two-year period between July 2002 and June 2004, more than 850 attacks by the LRA in Northern Uganda, causing at least 2,200 deaths and 3,200 abductions, were recorded (ICC 2005d). Thousands of adults and children have been killed in this conflict for no apparent reason at all, and there were reports that entire families were burnt to death as their villages were looted and destroyed. The LRA became notorious for abducting children, forcing boys to become child soldiers and girls to become sexual and labor slaves. By 2003, at least 1.6 million people had become internally displaced by the hostilities, and were forced to live in camps protected by the Ugandan military (ICC 2005d).

On December 16, 2003, the president of Uganda, Yoweri Museveni, referred the situation of crimes committed by the LRA since July 2002 to the ICC Prosecutor (ICC 2004a). Luis Moreno-Ocampo responded that any ICC investigation would be impartial and inclusive of all crimes committed in Northern Uganda that fall under ICC jurisdiction, whether they were carried out by the LRA or the Ugandan military (ICC 2005c). In July 2004, the Ugandan government conveyed their compliance with the ICC's position. That same month, ICC President Kirsch assigned the situation of Uganda to the PTC II, and Prosecutor Moreno-Ocampo opened an investigation of crimes committed in Northern Uganda (ICC 2004b, 2005a). The evidence that the investigative team uncovered indicated that the LRA was responsible for committing most of the crimes, so Moreno-Ocampo decided to focus on prosecuting LRA leaders (ICC 2005d).

On May 6, 2005, Moreno-Ocampo filed an application before the PTC II requesting that arrest warrants for crimes against humanity and war crimes be issued for five senior LRA commanders, including Joseph Kony (ICC 2005c, 2005d). The warrants were issued under seal by the PTC II on July 8th and were unsealed on October 13th. The LRA leader, Joseph Kony, was charged with twelve counts of crimes against humanity and twenty-one counts of war crimes, including murder, rape, enslavement, sexual enslavement, and the forced enlisting of children. The second in command of the LRA, Vincent Otti, was charged with eleven counts of crimes against humanity and twenty-one counts of war crimes. While one of the accused died in August 2006, the other four LRA commanders remained at large as of September 2011 (ICC 2011h).

Central African Republic

In January 2005, the ICC Prosecutor announced that he had received a request from the government of the Central African Republic (CAR) to initiate an investigation into crimes committed on the territory of that country (ICC 2005b). The fighting between government forces and rebel groups loyal to former President Ange-Felix Patasse began in 2002, when the army chief of staff François Bozizé ousted Patasse in a coup (Integrated Regional Information Networks 2005 [hereafter IRIN]). Since then, more than 150,000 people have become internally displaced within the CAR, mainly in the northern and eastern regions (UN News Service 2007). Tens of thousands of Central African refugees have fled to the neighboring countries of Chad and Cameroon (IRIN 2005, 2006).

In its referral letter, the CAR government revealed to the Prosecutor that while it had begun judicial proceedings concerning the most serious crimes, a decision had been reached in the country's highest judicial body, the *Cour de Cassation*, that the national justice system was unable to conduct adequately the investigations and prosecutions of those crimes (ICC 2007b). Thus, the CAR government suspended the proceedings and referred the crimes to the ICC. On May 22, 2007, the Prosecutor announced his decision to open an investigation in the CAR.

In May 2008, Moreno-Ocampo requested that the PTC III issue an arrest warrant for Jean-Pierre Bemba Gombo, the leader of the *Mouvement de libération du Congo* (Movement for the Liberation of Congo, MLC). Gombo was subsequently arrested by Belgian authorities later that month, and was transferred to ICC custody in July 2008. He was charged with two counts of crimes against humanity and three counts of war crimes in January 2009, and his trial began on November 22, 2010 (ICC 2011a).

Darfur, Sudan

The fourth situation that was referred to the ICC concerned the slaughter of civilians in the region of Darfur in western Sudan. In 2002, two rebel groups, the Sudan Liberation Movement/Army (SLM/A) and the Justice and Equality Movement (JEM), attacked government installations with the intent of launching an armed struggle to rectify the political and socio-economic marginalization of Darfur by the national government (International Commission of Inquiry on Darfur 2005 [hereafter ICID]). Since most of the Sudanese military was bogged down in a separate, protracted conflict in the southern region of the country, the government began recruiting, arming, and supporting the *Janjaweed*, militias composed mainly of nomadic Arab tribesmen who ride horses or camels. Due to the historical rivalry in Darfur between nomadic and sedentary tribes, it was easy for the government to persuade the *Janjaweed* to attack the rebel groups, whose members came predominantly from the sedentary tribes. The *Janjaweed* became notorious for engaging in indiscriminate murder, looting, and rape while raiding and destroying villages. By July 2007, it was estimated that at least 200,000 people had been killed in the conflict, and more than two million people had fled from their homes to refugee camps in Chad and the CAR (UNDPI 2007).

On September 18, 2004, the UN Security Council adopted Resolution 1564 under Chapter VII of the UN Charter, requesting Secretary-General Kofi Annan to appoint the ICID, which he did the following month. Following a three-month investigation, the ICID issued a report to the UN in January 2005, which concluded that the government of Sudan was not pursuing a policy of genocide in Darfur (ICID 2005). The report stressed, though, that the government and the *Janjaweed* were committing war crimes and crimes against humanity that were similar in gravity to genocide. Furthermore, the SLM/A and the JEM were also engaging in acts that constitute war crimes. Since the Sudanese judicial system was unwilling to investigate and prosecute those who were responsible for the crimes committed, the ICID report recommended that the situation of Darfur should be referred to the ICC. On March 31, 2005, the Security Council passed Resolution 1593, referring the case of crimes committed in Darfur since July 1, 2002 to the ICC Prosecutor (UNSG 2005).

As part of his preliminary examination, Prosecutor Moreno-Ocampo proceeded to collect thousands of documents on human rights violations in Darfur from various sources, including the Sudanese government, the ICID, the UN, the African Union, academic specialists, and the local and international media (ICC 2005e). More than fifty independent experts on the situation in Darfur were interviewed. Following a review of the evidence that was gathered, Moreno-Ocampo announced on June 6, 2005, that he would launch an investigation that would focus on the individuals who were the most responsible for crimes committed in Darfur.

On April 27, 2007, the PTC I issued arrest warrants for war crimes and crimes against humanity for both the former Sudanese Minister of State for the Interior, and current Minister of State for Humanitarian Affairs, Ahmad Muhammad Harun, and the *Janjaweed* commander Ali Muhammad Ali Abd-al-Rahman, also known as Ali Kushayb (ICC 2007c; UNDPI 2007). In 2009 and 2010, arrest warrants were issued for Sudanese President Omar Hassan Ahmad Al Bashir on five counts of crimes against humanity, two counts of war crimes, and three counts of genocide (ICC 2011b). All three suspects remain at large. But three other Sudanese suspects of war crimes have appeared before the ICC in 2009 and 2010 (ICC 2011h).

Kenya

For two months following the December 27, 2007 presidential election, Kenya suffered through ethnic violence which caused the deaths of around 1,220 people, injured more than 3,560, led to the rape of more than 900, and displaced around 350,000 (Jalloh 2011). Although Kenyan government officials promised to establish a special tribunal to prosecute the atrocity crimes, or else refer the situation to the ICC, Kenyan lawmakers twice voted down bills to create the tribunal. Due to Kenyan inaction regarding prosecutions, former UN Secretary-General Kofi Annan, who had helped broker the peace in Kenya, proceeded to transmit a list of suspected perpetrators of crimes against humanity to the ICC Prosecutor, according to the wishes of a Kenyan commission investigating the post-election violence.

On November 5, 2009, Prosecutor Moreno-Ocampo notified ICC President Song that he would submit a request to launch an investigation (ICC 2010). The following day, President Song assigned the situation to the PTC II, which authorized the Prosecutor on March 31, 2010 to begin an investigation *proprio motu* into the Kenyan situation. On March 8, 2011, the PTC II issued summons to six Kenyans, all but one with governmental positions, to appear before the ICC and be charged with crimes against humanity, including murder, forcible transfer of population, and persecution (ICC 2011h).

Libya

The Libyan civil war erupted on February 15, 2011, when initially peaceful protests against the Muammar Gaddafi regime were countered by military force. On February 26th, the UN Security Council decided unanimously to refer the situation in Libya to the ICC (ICC 2011g). Following a preliminary examination, the ICC Prosecutor decided on March 3rd to launch an investigation. On June 27th, the PTC I issued arrest warrants for Muammar Gaddafi, his son Saif Al-Islam Gaddafi, and the head of military intelligence Abdullah Al-Senussi, on the charges of crimes against humanity, including murder and persecution (ICC 2011e). The Gaddafi regime was toppled from power around one month later, and Muammar Gaddafi was located and killed by National Transitional Council troops on October 20, 2011. Nearly one month later, Saif Al-Islam Gaddafi and Abdullah Al-Senussi were both taken into custody by the new Libyan government.

Conclusion

The initiative to create an ICC demonstrates that middle powers are capable of skilled leadership on issues of human security. The middle powers brokered the strong and cohesive LMG, an amazing task in that the coalition included nearly one-third of all the countries in the world. Middle power officials assumed influential positions at the meetings of the Preparatory Committee and at the Rome Conference, which enabled them to control the negotiations on the ICC. Adriaan Bos from the Netherlands and Philippe Kirsch from Canada were astute chairmen, especially in their appointment of middle power delegates as issue coordinators. The LMG cultivated a close relationship with the CICC, thus forming a powerful, pro-ICC lobby at the Rome Conference. Most important, the LMG held firm when the United States and other members of the P5 sought to weaken certain provisions of the Rome Statute.

The Clinton administration was in favor of an ICC, as long as the United States was exempt from its jurisdiction. At the Rome Conference, the United States pushed unsuccessfully for UN Security Council oversight of the ICC, so that the United States would be able to veto any ICC investigation which conflicts with American interests. The Clinton administration was concerned that a fully independent ICC would conduct politically motivated trials of US military personnel without regard for their constitutional rights as American citizens. Although the United States tried to thwart the adoption of the Rome Statute, skilled diplomacy by the LMG managed to sidestep American opposition and established an ICC with a Prosecutor who can initiate investigations *proprio motu*.

President Clinton finally signed the Rome Statute in 2000 in order to give the United States a voice in the further development of the ICC, but President Bush withdrew the United States from the statute seventeen months later. It was not until the War in Iraq, and the ICC Prosecutor's decision that the numerous allegations of war crimes committed by the US-led coalition were inadmissible, that the Bush administration began to view the ICC as less of a threat to core American interests. The Obama administration has embarked on a wise policy of positive engagement with the ICC, which hopefully will lead to closer relations, and perhaps ICC membership, in the near future.

Notes

1 According to Benjamin Ferencz (1980), the plan to set up an International Prize Court was ultimately rejected due to the apprehension of several major powers about the uncertainty of the rules of international law. The Naval Conference in London (December 1908-February 1909) produced a Code of Naval Law covering blockade, contraband, the limits of permissible search, and the destruction of prizes from neutral parties. But the Prize Court bill was rejected by the British House of Lords because food imports were listed as "contraband" items, and thus subject to seizure by an adversary (e.g. Germany) during war. Since the Code was not accepted, not a single state ratified the Hague Convention XII.

2 According to Herman Von Hebel (1999), the General Assembly merely postponed issues. On December 4, 1954, General Assembly Resolution 9/895 set up a second Committee to Define Aggression, which was mandated to issue a report in 1956. That same day, General Assembly Resolution 9/897 delayed

further discussion of the draft Code of Offences until a decision could be reached on the definition of aggression.

3 In June 2000, the Steering Committee consisted of Amnesty International, *Asociacion pro Derechos Humanos* (Association for Human Rights, APRODEH), the European Law Students Association, *Fédération International des Ligues des Droits de l'Homme* (International Federation of Leagues of Human Rights, FIDH), HRW, the International Center for Human Rights and Democratic Development (Rights and Democracy), the International Commission of Jurists, the Lawyers Committee for Human Rights, No Peace Without Justice, Parliamentarians for Global Action (PGA), the Women's Caucus for Gender Justice, and the WFM. See Pace and Schense 2001, fn. 7.

4 According to William Pace and Jennifer Schense, the CICC had established national networks in twenty-six countries by June 2000, including Argentina, Bangladesh, Belgium, Brazil, Burundi, Cameroon, Canada, Chile, Colombia, Egypt, France, Germany, Ghana, Japan, Kenya, Mexico, Peru, Poland, the Russian Federation, Senegal, South Africa, Spain, Thailand, the United Kingdom, the United States, and Venezuela. See Pace and Schense 2001, fn. 8.

5 The members were Andorra, Argentina, Australia, Austria, Belgium, Benin, Bosnia and Herzegovina, Brazil, Brunei Darussalam, Bulgaria, Burkina Faso, Burundi, Canada, Chile, Congo, Costa Rica, Côte d'Ivoire, Croatia, the Czech Republic, Denmark, Egypt, Estonia, Fiji, Finland, Gabon, Georgia, Germany, Ghana, Greece, Hungary, Iceland, Ireland, Italy, Jordan, Latvia, Lesotho, Liechtenstein, Lithuania, Luxembourg, Malawi, Malta, Namibia, the Netherlands, New Zealand, Norway, the Philippines, Poland, Portugal, the Republic of Korea, Romania, Samoa, San Marino, Senegal, Sierra Leone, Slovakia, Slovenia, the Solomon Islands, South Africa, Spain, Swaziland, Sweden, Switzerland, Trinidad and Tobago, the United Kingdom, Venezuela, Zambia, and Zimbabwe. See Pace and Schense 2001, fn. 5.

6 Another middle power, Australia, would later succeed Canada as the leader of the LMG in Rome. See Pace 1999.

7 On June 23, 2004, the Bush administration withdrew its request to renew the 2003 Security Council Resolution 1487, which would have extended the exemption for another year, after it realized that the other Security Council members would not vote in favor, due to anger over the abuse of Iraqi prisoners by American soldiers during the war in Iraq. See Aldinger 2004 and AMICC 2004.

8 The Nethercutt Amendment permitted presidential exemptions for NATO members, major non-NATO allies of the United States, and Millennium Fund countries. See AMICC 2002b.

CHAPTER SIX

Regulating the legal trade in small arms and light weapons

The previous chapters described three cases where middle power leadership has been successful in achieving human security initiatives. The middle powers were instrumental in forming SHIRBRIG, establishing the Ottawa Convention, and creating the ICC. They had to deal with a US foreign policy that expressed indifference in the case of SHIRBRIG, acquiescence with regards to the APL ban, and hostility concerning the ICC.

The effectiveness of middle power leadership on human security is evaluated further in this chapter, through an analysis of the middle powers' attempt to achieve international restrictions on the legal trade in small arms and light weapons (SALW). Unfortunately, in sharp contrast to the previous cases, the SALW initiative has failed to fulfill its objectives. This chapter examines why.

I begin with a review of alternative definitions of SALW, the impact of the global proliferation of SALW, and the magnitude of the SALW industry. This is followed by a discussion of the political and cultural relevance of small arms to Americans. I then analyze the middle-power-led campaign to regulate the licit trade in SALW, and highlight why it has not been successful so far.

The problematic proliferation of small arms and light weapons

What are "small arms and light weapons"?

There is considerable disagreement as to which weapons may be classified as SALW. Andrew Latham (1996) described four different definitions of light weapons which are in current use. First, light weapons have been defined as those weapons which are not included in existing data collections on major weapons, such as the UN Register of Conventional Arms and the annual register of major weapons transfers published by the Stockholm International Peace Research Institute (SIPRI). A second definition of light weapons refers to weapons carried by infantry, such as pistols, grenade launchers, and light rocket launchers. But this definition excludes many weapons which are not covered by the UN and SIPRI registers, such as antiaircraft artillery and heavy machine guns.

A third definition considers light weapons as those transportable by light vehicles or animals, including heavy machine guns, rifles, mortars, and some artillery. The problem with this definition is that it does not distinguish clearly enough between light weapons and major conventional weapons systems. Nevertheless, in 1983, NATO adopted a similar definition of light weapons as "all crew-portable direct fire weapons of a caliber less than 50mm [with] a secondary capability to defeat light armor and helicopters" (Latham 1996, 2). Finally, light weapons have been defined as the weapons used in intrastate conflict, a broad definition which may encompass anything from firearms to aircraft. Taking the shortcomings of these definitions into consideration, Andrew Latham adopted a broad conceptualization of light weapons as:

> All armaments that fall below the threshold of major conventional weapons systems (which are understood to include those weapons encompassed by the seven categories of the United Nations Register of Conventional Arms: battle tanks, armored combat vehicles, large caliber artillery, combat aircraft, attack helicopters, warships, and missiles/launchers). (Latham 1996, 3)

This definition includes such weapons as assault rifles, machine guns, light antitank weapons, light mortars, shoulder-fired antiaircraft missiles, and landmines. The United States' definition of SALW has been expressed by Under-Secretary of State John R. Bolton as:

> The strictly military arms—automatic rifles, machine guns, shoulder-fired missile and rocket systems, light mortars—that are contributing to continued violence and suffering in regions of conflict around the world. We separate these military arms from firearms such as hunting rifles and pistols, which are commonly owned and used by citizens in many countries. ("UN Conference on Illicit Trade in Small Arms" 2001, 902 [hereafter "UN Conference"])

In a 1997 report, the UN Panel of Governmental Experts on Small Arms presented its own classification of SALW (Garcia 2002). According to the Panel, "small arms" include revolvers, self-loading pistols, rifles, carbines, sub-machine guns, light machine guns, and assault rifles. The category of "light weapons" comprises heavy machine guns, portable antiaircraft and antitank guns, recoilless rifles, hand-held under-barrel and mounted grenade launchers, portable launchers of antiaircraft and antitank missile and rocket systems, and mortars of less than 100 mm caliber. Cartridges for small arms, shells and missiles for light weapons, landmines, and antitank and antipersonnel grenades are classified under a third category of "ammunition and explosives."

One of the main reasons why the issue of regulating the trade in small arms and light weapons has been contentious is because there is no consensus as to which types of weapons should be classified as SALW. According to the broadest definition, SALW includes not only firearms and portable weapons, but landmines, armor, and aircraft as well. A minimalist definition of SALW is the one held by the US government, which excludes firearms that are not designed for military purposes (even though they could be used for killing people).

The devastating impact of SALW

During the decade after the Cold War ended, at least four million people were killed by SALW in armed conflicts ("Under the Gun" 2001). Around ninety percent of these victims were civilians, and eighty percent were women and children. Jayantha Dhanapala, the former UN Under-Secretary-General for Disarmament Affairs (1998–2003), emphasized that at least half a million people each year die from SALW, including around 300,000 from armed conflict and approximately 200,000 from homicides and suicides (Dhanapala 2002). Drawing on data compiled in 2004 by the Small Arms Survey, a research group based in Geneva, the International Action Network on Small Arms (IANSA) claimed that firearms cause more than 300,000 deaths and one million injuries annually (IANSA 2006). Between 60,000 and 90,000 people are killed by firearms in the context of armed conflict. More than 200,000 people become victims of homicide and around 50,000 people use a gun to commit suicide each year. The use of SALW in intrastate conflicts has also displaced millions of people. According to Michael Klare and Robert Rotberg, over twenty-two million people have become refugees due to the use of SALW in civil wars, including eight million in Africa (Klare and Rotberg 1999). The proliferation of SALW has clearly been detrimental for human security.

Klare and Rotberg stressed that "more people have been killed by small arms and light weapons in recent wars than by major weapons systems" (Klare and Rotberg 1999, 7). But as Michael Renner indicated, "small arms are the orphans of arms control" (Renner 1999, 22). During the Cold War, arms control negotiations focused on major weapons systems, hence, the global community failed to adopt international norms regarding the production, transfer, and possession of SALW. The 1990s witnessed a major transformation in the nature of conflict, however, as traditional warfare between nation-states was largely supplanted by intrastate conflict between ethnic and sectarian groups (Boutwell and Klare 1999). Due to their accessibility, low cost, and portability, SALW are the preferred weapons of combatants in intrastate conflicts. It is far easier for guerrillas, militias, drug traffickers, and terrorist groups to acquire and use SALW than major weapons systems (Klare 1999).

The adverse consequences of the proliferation of SALW have been expressed eloquently by Klare and Rotberg:

> Not every massacre has resulted from the easy availability of small arms. Cause and effect is impossible to establish. But in every recent case of large-scale mayhem, intercommunal conflict, ethnic or religious hostility, and racial violence in the developing world, small arms have been used to increase the scale and carnage of the fighting. Absent AK-47s or Uzis, inexpensive and universally accessible, intercommunal combat would have been harder to mount, genocidal instincts more difficult to fuel, and conflicts over perceived differences and competition for resources much less destructive. The impoverishment and immiseration of much of the developing world cannot be ascribed solely either to war or to the ease of acquiring small arms. But the destructive quality of small arms, and their ubiquity, has hardly eased efforts of economic development. (Klare and Rotberg 1999, 7–8)

Jayantha Dhanapala concurred that the proliferation of SALW has been culpable for the rise in violence in many countries:

> Although the widespread availability of these weapons alone does not cause war, in many situations their accumulation becomes excessive and destabilizing. According to the highly debated "accessibility thesis," the widespread availability of small arms and light weapons leads to increased levels of violence. Although there is no conclusive proof that guns cause violence—gun-owner advocates argue against it—there is ample evidence that the proliferation of weapons is closely associated with levels of violence. (Dhanapala 2002, 163–4)

According to Andrew Latham (1996), there are five interrelated problems due to the diffusion of light weapons. First, both traditional and modern institutions of human security are undermined by easy access to light weapons. Subnational groups with grievances may use SALW against persons, communities, or institutions of governance, public order, or national defense. Second, the accumulation of light weapons helps generate a culture of violence. Societies that are awash in SALW and suffer protracted conflict may become culturally militarized, where violent strategies for resolving societal problems become routine. Third, facilitating the acquisition of these weapons helps prop up authoritarian regimes and hinders the development of democratic institutions. Elite groups which control the state may use SALW to prevent the emergence of more pluralist or representative politics. Fourth, some types of light weapons, such as antipersonnel landmines and fuel-air explosives, are particularly inhumane, because they strike both military targets and civilians without discrimination. Finally, the use of certain types of light weapons can be detrimental for postconflict efforts at peace-building and economic reconstruction. A good example is the unrecorded laying of antipersonnel landmines, which poses severe hazards for the postwar task of mine clearance, and prevents the cultivation of arable land for agriculture.

Michael Klare (1999) summarized the major findings of several studies on the basic dynamics of the SALW trade, and the implications for global peace and security. First, there is a close and symbiotic relationship between trafficking in SALW and contemporary forms of violent conflict. The intrastate conflicts of the post-Cold War era tend to be fought primarily with SALW, since these weapons are easy to procure and operate. Second, the initiation of internal conflict in weak and divided societies often generates a SALW arms race between subnational groups. While government forces have access to legitimate suppliers of SALW, insurgent groups often obtain these weapons through illicit channels. Third, outbreaks of intrastate conflict and internal arms races are fostered by an overabundance of SALW worldwide. During the Cold War, the superpowers produced and distributed vast quantities of SALW to their allies. These surplus weapons are now being redistributed globally through both licit and illicit markets. In addition, new supplies of SALW are being manufactured in dozens of countries, as the SALW technology has spread to developing states. Fourth, even relatively small amounts of SALW can have highly destabilizing effects in societies that are divided along sectarian lines. The acquisition of SALW by extremist groups, ethnic militias, and criminal gangs has triggered massacres in vulnerable societies. Finally, there are multiple channels, public and private, licit and illicit, through which parties in conflict can obtain SALW, a subject which shall be discussed in the next section.

The global SALW industry

Unfortunately, there is a scarcity of data on the worldwide production and stockpiling of SALW. The UN has estimated that there are at least five hundred million SALW in global circulation, which amounts to approximately one for every twelve people ("Under the Gun" 2001). Michael Klare and Robert Rotberg (1999) claimed that by the end of the 1990s, the number of military-style small arms in the developing world lay somewhere between one and five hundred million, a significant increase from the figure of forty million in 1990. There were also hundreds of millions of civilian-type small arms, including handguns and rifles, around the world.

Based on data gathered from thirty-three participating countries, the *United Nations International Study on Firearm Regulation* (1998) reported that around thirty-four million firearms were owned by civilians. Michael Renner (1999) cautioned that this is a very conservative figure, considering that the number of registered firearms may be a minority of the total amount. For instance, in 1999, there were 3.5 million registered civilian small arms in South Africa, but perhaps as many as five to eight million unregistered weapons. That same year, Canada had a grand total of twenty-one to twenty-five million firearms, of which only seven million were registered. According to a 2006 IANSA publication, civilians own 59 percent of the total number of firearms in the world, while government armed forces possess 37.9 percent, police forces have 2.8 percent, and armed groups hold 0.2 percent (IANSA 2006).

SALW are manufactured in around seventy countries, including nineteen in the developing world. The major producers and exporters of SALW include Austria, Belgium, Brazil, Bulgaria, China, the Czech Republic, Egypt, France, Germany, Israel, Italy, Russia, Singapore, South Africa, South Korea, Switzerland, the United Kingdom, and the United States (Klare and Rotberg 1999; Renner 1999). Some of the leading firms in the military and civilian small arms industry are Beretta (Italy), *Fabrique Nationale Herstal* (Belgium), Heckler & Koch (Germany), Israeli Military Industries, *Schweizerische Industrie Gesellschaft* (Switzerland), and Steyr-Daimler-Puch (Austria). The SALW industry is quite lucrative; in 1996 alone, the US State and Commerce Departments approved $530 million worth of small arms exports by private firms (Renner 1999). Data on US production and exports of small arms and light weapons may be the best available measures of the profitability of the global SALW industry. Lora Lumpe of the International Peace Research Institute in Oslo made the point that:

> because other governments are not open about their light weapons shipments, it is not possible to rank the leading sellers in the global small arms trade; however, given the sheer magnitude of U.S. licenses and sales . . . it is reasonable to assume that the United States dominates the small arms market just as it dominates the market for larger weapons systems. (Lumpe 1999, 27)

In fact, the United States is the only country to issue annual statistics on SALW exports, through reports to Congress. Michael Klare and Robert Rotberg highlighted the fact that "even those countries, like Belgium, that are officially anxious to reduce the spread of small arms, shield their own manufacturing industries by refusing to release information on numbers and destinations. That is a common pattern" (Klare and Rotberg 1999, 9).

There is also little data on the global trade in SALW, estimated to be worth around seven billion dollars annually (Klare and Rotberg 1999). As Michael Renner argued, "many analysts . . . believe that the demand for and trade in small weapons continues to be robust and may even be accelerating—in marked contrast to the plummeting trade in major weapons systems since the late 1980s" (Renner 1999, 22). According to figures from the US Arms Control and Disarmament Agency (ACDA), the international trade in small arms and ammunition accounts for around thirteen percent of the total conventional arms trade, or around three billion dollars per year (Renner 1999). But small arms expert Michael Klare argued that ACDA's estimate is too low because it excludes transfers of machine guns, light artillery, and antitank weapons, all of which may be considered as light weapons. The Small Arms Survey has calculated that between ten and twenty percent of the worldwide trade in SALW, valued at approximately one billion dollars per year, is made up of illegal transactions ("Big Damage; Small Arms" 2001). But the situation may be even worse, according to the UN's estimates that only fifty to sixty percent of the SALW trade is legal ("Under the Gun" 2001).

Many of the weapons that are originally sold through legal channels eventually end up on the black market. Supplies of SALW are frequently stolen or captured from national military forces by insurgent, terrorist, or criminal groups. With the denouement of the Cold War era and the proxy wars fought between the United States and the Soviet Union in less developed countries, much of the developing world was left with huge stockpiles of SALW, including Afghanistan (around ten million weapons), West Africa (seven million), and Central America (two million). These stockpiles may be particularly vulnerable to theft by criminal and insurgent groups, who could then sell the stolen SALW on the black market ("Under the Gun" 2001).

Michael Klare (1999) discussed the various means through which actors may acquire SALW. The first is government-to-government transfers, where one government either sells or donates SALW to another government via overt, legal channels. These sorts of transfers were far more common during the Cold War than in the present era. The second, and most common, means of transferring SALW is through government-sanctioned commercial sales. Private firms adhere to government export regulations in selling SALW to a state or commercial entity that is approved by their government.

A third way in which SALW is diffused is through covert or "grey market" operations. Government agencies or government-backed private firms may sell or donate SALW to illicit recipients in another country in order to achieve political or strategic objectives. These types of transactions were quite common during the Cold War, when the United States and the Soviet Union provided SALW to insurgent groups that were seeking to overthrow regimes that were allies of the other superpower. Transfers of this nature still occur in the contemporary period, as some governments continue to aid insurgent groups. Furthermore, government agencies may engage in the covert transfer of SALW to domestic militias or death squads who serve the interests of the government.

Finally, SALW may be transferred through black-market transactions or theft. Private actors may engineer the covert sale of SALW which have been procured illicitly, in clear violation of national laws. Insurgent groups, ethnic militias, terrorist organizations, and warlords usually acquire SALW through the black market, since legal trade channels are closed for them. These actors also rely on theft from government stockpiles, or clandestine collaborations with corrupt military officials who are willing to risk incarceration in order to make high profits from selling SALW to enemies of the state.

The political and cultural importance of small arms to the United States

The Second Amendment to the US Constitution

The issue of small arms has always been central to American politics. According to the Second Amendment to the US Constitution, "a well regulated Militia, being necessary to the security of a free State, the right of the people to keep and bear Arms, shall not be infringed" ("The Constitution of the United States of America" [1787] 2002). Sanford Levinson remarked that of all the provisions in the Constitution, the Second Amendment is "perhaps one of the worst drafted" (Levinson 1989, 643). Karen O'Connor and Graham Barron (1998) added that the Second Amendment has had little judicial interpretation, leaving the question open as to whether or not the right to bear arms is tied to or based on membership in a militia. Lee Kennett and James LaVerne Anderson (1975) suggested that the constitutional right to bear arms stems from the need to maintain a militia for national defense:

> The judicial interpretations of the Second Amendment do not clearly define the meaning of the phrase "the right to bear arms." There are no decisions supporting the unlimited right of the people to bear arms. The right has usually been viewed in terms of maintenance of the militia and the responsibility of citizens to defend the country. (Kennett and Anderson 1975, 78)

Moreover, it is debatable whether the term "well-regulated militia" restricts the right to bear arms solely to members of government-sponsored militias, or permits members of informal militia groups to bear arms as well (O'Connor and Barron 1998). William Weir (1997) argued that the meaning of "militia" has only changed slightly over two centuries. While the militia included all male citizens between the ages of sixteen and sixty at the time the Bill of Rights was written, under the current United States Code, the militia consists of all able-bodied males at least seventeen and under forty-five years of age. Nevertheless, the intent of the Second Amendment is timeless: "because the militia . . . have weapons, they are the greatest bulwark against foreign invasion or domestic usurpation. Therefore they shall not be disarmed" (Weir 1997, 34).

In order to understand the logic behind the Second Amendment, some knowledge of early American history is useful. In England, political philosophers challenged the concept of a standing army, because it could be used to crack down on the liberties of the citizenry, and endorsed the idea of citizen-soldiers instead. According to Lee Kennett and James LaVerne Anderson, "reliance on citizen-soldiers, as opposed to a professional army, became the hallmark of the emerging American national conscience" (Kennett and Anderson 1975, 61). The original settlers in America also distrusted the English standing armies, which they felt were infringing upon their personal liberties (O'Connor and Barron 1998). In the Declaration of Independence, Thomas Jefferson objected to King George III's maintenance of standing armies in times of peace, denial of civilian control over the military, quartering of troops, and use of foreign mercenaries against the Americans ("The Declaration of Independence" [1776] 2002). Most of the American colonies required by law that nearly all white men carry arms in local militias, and the American Revolution was fought by men in state militias, which were mobilized with the objective of defending the colonies (O'Connor and Barron 1998).

Following the Revolutionary War, the Federalists tried to convince the Anti-Federalists that a standing army of twenty-five to thirty thousand soldiers, controlled by the federal government, would not be able to suppress the liberties of people, due to the presence of a citizen militia numbering nearly half a million armed men (Weir 1997). During the ratification conventions for the Constitution, there were many demands for the inclusion of an amendment in the future Bill of Rights which would guarantee the right of citizens to bear arms. Karen O'Connor and Graham Barron suggested that these appeals "were more numerous than demands for explicit protection of the right to free speech or assembly" (O'Connor and Barron 1998, 76). In response, James Madison wrote the Second Amendment, with the intention of protecting the arms of the citizenry so that they would not fear the creation of a new national government.

The American gun culture

It can be argued that there is no longer a need for armed civilian militias in the contemporary era. American national defense is firmly in the hands of the overwhelmingly powerful United States military. Furthermore, the record of the past two centuries has shown that the US military is not a threat to the liberties of American citizens. It should also be noted that, contrary to the vision of Madison, it would be a futile task nowadays for citizen militias to attempt to mount a resistance to the US military. Thus, the traditional justification for the right of American citizens to bear arms is no longer valid. But Americans still remain adamant about defending their Second Amendment rights in the twenty-first century.

The explanation for this behavior lies in the American gun culture. Robert Spitzer (1995) argued that many Americans have a deep sentimental attachment to firearms, based on three factors: the presence and proliferation of firearms in the United States since the colonial period, the connection between personal ownership of weapons and the United States' frontier legacy, and the cultural mythology about guns, which is disseminated through the media. Spitzer suggested that the American gun culture contains at least two elements which have persisted since the colonial era. First, the "hunting/sporting ethos" dates back to when the United States was an agrarian, subsistence nation. American settlers needed to hunt game for their very survival. Moreover, the fur market encouraged hunting and trapping as a source of income. The acquisition of shooting skills was seen as a rite of passage for teenage boys. Competitive shooting was a popular sport at that time, as such competitions helped sharpshooters hone their skills. Due to the hunting heritage, around fourteen million Americans identify themselves as hunters in the present period.

Second, a "militia/frontier ethos" developed early in American history. Able-bodied male settlers had the responsibility of being citizen-soldiers, since no full-time army existed. The young colonies were vulnerable to attacks from foreign armies and Native American tribes, and their survival depended on the manpower and weaponry of the citizen militias. According to Robert Spitzer, "the death knell of the citizen militia was its abysmal performance in the War of 1812, after which time it ceased to play any active role in national defense. Despite this fact, the militia tradition has survived" (Spitzer 1995, 10). As the settlers moved westward, a frontier legacy grew. Settlers armed themselves with Colts, Remingtons, Smith and Wessons, and Winchesters in order to deal with outlaws and Native American warriors. Tragically, the proliferation of firearms triggered many massacres of Native Americans as white settlements spread in the Western United States.

But the image of the United States as a "gunfighter nation" is a myth. Robert Spitzer argued that "the so-called taming of the West was in fact an agricultural and commercial movement, attributable primarily to ranchers and farmers, not gun-slinging cowboys" (Spitzer 1995, 10). Hollywood films, and the entertainment industry in general, have romanticized the frontier legacy and exaggerated the contribution of firearms to the settling of the West. As Gregg Lee Carter stated:

> In reality, though stories of frontier violence were so popular, the level of violence—especially gun violence—in non-frontier America was but a fraction of that actually occurring on the frontier, which itself was

considerably less than that depicted in the Wild West shows and dime novels. (Carter 1997, 42)

In summary, the gun culture in the United States fueled by both a hunting/sporting ethos and a militia/frontier ethos, has influenced American citizens to defend their constitutional right, under the Second Amendment, to bear arms. Their belief in the necessity of the Second Amendment has outlived the original justifications for the amendment: the need to form citizen militias for the defense of the nation and the protection of personal liberties. The Second Amendment is still regarded as sacred in the contemporary period, as was demonstrated when the US government invoked it when blocking an attempt by the international community to adopt restrictions on the legal trade in SALW. The chapter will now turn to a discussion of this initiative, and the US reaction to it, in the following section.

The initiative to regulate the legal trade in SALW

The campaign is launched

The issue of SALW was first placed on the international agenda in October 1993, when President Alpha Oumar Konare of Mali, a country severely affected by the illicit influx of small arms, made a request to the UN Secretary-General for assistance with the problem of SALW proliferation in West Africa (NGO Committee on Disarmament 2001d; Smaldone 1999). A UN Advisory Mission visited Mali in August 1994, as well as Burkina Faso, Chad, Côte d'Ivoire, Mauritania, Niger, and Senegal in February and March of 1995. The mission reported that insecurity at the levels of the individual, locality, nation, and subregion was hindering the socioeconomic development of these countries, and fueling the demand for firearms. Furthermore, the mission adviced that in order to curb the spread of SALW, the UNDP should adopt a subregional approach for cultivating human security and good governance.

In response to the mission's recommendations, Mali hosted a UNIDIR/UNDP conference on Conflict Prevention, Disarmament, and Development in West Africa from November 25–29, 1996. At the conference, Mali's foreign minister proposed a subregional moratorium on the import, export, and manufacture of light weapons. Nearly two years later, in October 1998, the sixteen members of ECOWAS signed a Moratorium on the Exportation, Importation, and Manufacture of Light Weapons. The moratorium has some shortcomings: it is a politically, but not legally, binding agreement that is renewable for periods of three years, and it was implemented with few operational guidelines and limited technical and financial support (Meek 2000). But the moratorium was the first regional initiative to curtail the proliferation of SALW in West Africa.

Other regional organizations have also taken action. During its presidency of the EU in June 1997, the Netherlands negotiated the adoption of a Program for Preventing and Combating Illicit Trafficking in Conventional Arms. On May 25, 1998, the EU agreed to a Code of Conduct on Arms Exports, which requires member states to halt arms transfers that would likely be used for internal repression in states with dubious human rights records, as well as arms transfers that may provoke or prolong conflict in a particular area of tension (Klare and Rotberg 1999). But Paul Eavis and William Benson indicated that, despite these initiatives on both the illicit trafficking and the licit sales of SALW, "relatively little has been done in practice to specifically target and prevent the export (both legal and illicit) of light weapons from the EU" (Eavis and Benson 1999, 99).

On November 14, 1997, the members of the OAS signed the Inter-American Convention Against the Illicit Manufacturing of and Trafficking in Firearms, Ammunition, Explosives and Other Related Materials. The convention, which entered into force in 1998, requires states parties to implement national legislation making the illicit production and transfer of firearms criminal offenses, and necessitates that firearms be marked so as to make them traceable if they are diverted into illicit channels (Klare and Rotberg 1999). As of September 2011, thirty-four states had signed or acceded to the treaty, but four signatories—Canada, Jamaica, Saint Vincent and the Grenadines, and the United States—had yet to ratify it (OAS 2011).

The Wassenaar Arrangement on Export Controls for Conventional Arms and Dual-Use Goods and Technologies was adopted in December 1995 by thirty-five industrialized countries, including the United States As of November 2011, forty states were participating (Wassenaar Arrangement 2011). The arrangement promotes transparency and responsibility among member states to ensure that their transfers of conventional arms and dual-use goods and technologies do not produce destabilizing accumulations. But decisions whether or not to transfer weapons are made solely by each participating state. The Wassenaar Arrangement serves only as a forum for the exchange of information on weapons sales, with no enforcement capability (Klare and Rotberg 1999). Moreover, the arrangement does not include small arms explicitly.

The middle powers and the United Nations SALW conference

Apprehensive about the growing risk to UN peacekeepers from the spread of SALW, the UN General Assembly passed a resolution on *Assistance to*

States for curbing the illicit traffic in small arms and collecting them for the first time in 1994, which invited member states to adopt national control measures and requested international support for their efforts (NGO Committee on Disarmament 2001d). The following year, based on the recommendations of a UN report, Japan sponsored a General Assembly resolution which called for the creation of a Panel of Governmental Experts on Small Arms to study the nature of the SALW problem and possible solutions (Lozano 1999). On August 27, 1997, the Panel issued its report, which made suggestions concerning the safeguarding of SALW, as well as measures to prevent and reduce the destabilizing effects from the excessive stockpiling and transfers of these weapons. The Panel also called for the convening of an international conference on the illicit trade in SALW in all its aspects. In response, on December 9, 1997, General Assembly Resolution 52/38 J, *Small arms*, endorsed the Panel's recommendations. The resolution also authorized the Secretary-General to begin planning for an international conference on SALW, and to set up a twenty-three member Group of Governmental Experts on Small Arms that would report on progress in the implementation of the Panel's recommendations.

A parallel initiative to combat the illegal trade in firearms was underway within the Commission on Crime Prevention and Criminal Justice (CCPCJ), a subsidiary body of ECOSOC. Following the release of a study of international firearms regulations that was commissioned by the CCPCJ three years earlier, the CCPCJ passed a resolution on April 28, 1998 which requested that the UN General Assembly produce an international instrument to combat the illicit production and trade in firearms, their components, and ammunition, and suggested that such an instrument could be modeled on the Inter-American Firearms Convention (Lozano 1999).

On May 31, 2001, the General Assembly adopted the Protocol against the Illicit Manufacturing of and Trafficking in Firearms, Their Parts and Components and Ammunition, supplementing the United Nations Convention against Transnational Organized Crime (UNGA 2001). The Firearms Protocol requires states parties to ensure that firearms are marked for identification at their time of manufacture so as to render them traceable; maintain records of their international sales of firearms, components, and ammunition for at least ten years; and cooperate with other states parties and international organizations by sharing the information, training, and technical assistance necessary for the eradication of the illegal manufacturing of and trade in firearms. The protocol entered into force on July 3, 2005, ninety days after its fortieth ratification. As of September 2011, eighty-nine states had become parties to the Firearms Protocol, but the United States had never signed or acceded to it (UN Treaty Collection 2011). The Firearms Protocol is currently the sole legally binding instrument to regulate small arms at the global level (UN Program of Action Implementation Support System 2011).

In its September 1999 report, the Group of Governmental Experts on Small Arms recommended that an international conference on the illicit SALW trade should focus on those small arms and light weapons which are manufactured to military specifications, thus, rifles and firearms used for hunting and sports should be excluded from consideration. In December 1999, the General Assembly called for the convening of a UN Conference on the Illicit Trade in Small Arms and Light Weapons in All Its Aspects, and created a Preparatory Committee, which began work in February 2001 (NGO Committee on Disarmament 2001d).

Canada expressed its view that both the Preparatory Committee and the international conference should address all issues related to the excessive accumulation and uncontrolled proliferation of SALW, not merely the problem of illicit transfers (UNSG 1999, 2000a). On July 21, 2001, Canada submitted a working paper to the Preparatory Committee, which offered suggestions regarding the format and contents of an action plan on SALW. Among its many recommendations, Canada proposed that the plan should examine the relationship between the licit and illicit aspects of the SALW problem, and suggested that states should make a commitment "to exercise the maximum practicable restraint on the legal manufacture and transfer of small arms and to enhance efforts to prevent the illicit manufacture and transfer of such weapons" (UN Preparatory Committee for the UN Conference on the Illicit Trade in SALW in All Its Aspects 2000, 3 [hereafter UN Preparatory Committee]). Canada also stressed that the work of the Preparatory Committee should complement, not duplicate, the contents of the UN Firearms Protocol.

But from the first session of the Preparatory Committee, the Bill Clinton administration warned that it would not accept any legally binding international treaty that either constrains the legitimate trade in SALW by US nationals (including sales to nonstate actors), or infringes on the constitutional right of American citizens, under the Second Amendment, to own firearms ("UN Conference" 2001). The United States was committed, however, to the elimination of the illicit trade in military-type SALW. The objectives of US arms transfer policy included both the prevention of arms transfers which may destabilize or threaten regional peace and security, and the promotion of national and multilateral responsibility, restraint, and transparency in the arms trade (US Department of State 1998). The Clinton administration did make it clear during the meetings of the Preparatory Committee that it would accept the establishment of a program of action designed to curb the illicit SALW trade across international borders, a position which the George W. Bush administration would adopt as well ("UN Conference" 2001).

Similar to other human security initiatives, the like-minded states forged a close working relationship with NGOs on the SALW campaign. The International Action Network on Small Arms, a coalition of pro-regulation NGOs around the world, was created following meetings of NGO representatives in Orillia, Ontario, Canada in August 1998, and Brussels, Belgium in October 1998 (Klare and Rotberg 1999). The objectives of IANSA extend beyond the regulation of the licit SALW trade and the eradication of illicit transfers, to include the elimination of cultures of violence and the removal of SALW from postconflict societies (Clegg 1999).

But the NGOs realized from the start that they would be unable to push for a total ban on SALW in the same manner as they helped achieve a ban on antipersonnel landmines, for two reasons (Clegg 1999). First, the fact that civilian ownership of small arms is legal in countries worldwide means that there is no universal acceptance of the need to ban such weapons. Second, most people would concur that light weapons do have some legitimate uses, such as for peacekeeping operations.

Since many states are manufacturers and exporters of SALW, pro-regulation NGOs are concerned that governmental action on any SALW initiative would be overly cautious and incremental. While pro-regulation NGOs have emphasized the need to regulate the legal trade in SALW, most governments have only been willing to address the illicit trade. Liz Clegg (1999) described why national governments have been hesitant to take action on the licit SALW trade. First, sales and direct transfers of SALW to allies have been influential foreign policy tools for the governments of SALW-exporting countries, and have considerable domestic economic benefits as well. Second, in order to remain competitive in the global SALW market, domestic arms manufacturers campaign against the implementation of national and international restrictions on the legal trade. Finally, powerful gun lobbies, such as the National Rifle Association of America (NRA), mobilize against any legislation that would restrict firearms ownership.

The middle powers organized several meetings and workshops leading up to the UN conference on the illicit trade in SALW (UN 2001). In July 1998, Norway hosted the Oslo Meeting on Small Arms, which produced "Elements of a Common Understanding" on the need for both immediate action to prevent the illicit transfer of SALW as well as tighter controls on legal transfers (Clegg 1999; The Oslo Meeting on Small Arms 1998). A follow-up meeting (Oslo II) was held in December 1999. At the "Sustainable Disarmament for Sustainable Development" Brussels conference in October 1998 (which coincided with the NGO meeting), ninety-eight governments announced a "Brussels Call for Action" on light weapons.

The government of Canada co-organized regional SALW conferences and seminars together with Sri Lanka (June 2000), Poland (September 2000 and September 2001), Bulgaria (October 2000), Cambodia and Japan (February 2001), Hungary (April 2001), and the European Union (May 2001, under the Swedish Presidency of the EU). On November 7, 2000, the Canadian Joint Delegation to NATO and the Center for European Security and Disarmament convened a roundtable on Small Arms and Europe-Atlantic Security at NATO headquarters in Brussels. Canada also hosted an OAS seminar on the illicit SALW trade in Ottawa in May 2001.

The governments of the Netherlands and Hungary cooperated in organizing an expert workshop on the destruction of SALW as an aspect of stockpile management and weapons collection in postconflict situations, which was held in The Hague in September 2000. That same year, the London-based NGO Saferworld co-hosted three different seminars on SALW, together with the foreign affairs ministries of Poland, the Czech Republic, and Hungary, respectively. In addition, the Human Security Network discussed the topic of the SALW trade at its Second Ministerial Meeting in Lucerne, Switzerland in May 2000.

The New York conference on SALW

The United Nations Conference on the Illicit Trade in Small Arms and Light Weapons in All Its Aspects was held in New York City from July 9th to 20th, 2001 (Humanitarian Coalition on Small Arms 2001; Klare 2001; NGO Committee on Disarmament 2001a, 2001c; UN 2001; "UN Conference" 2001). Representatives from more than 140 states and over forty NGOs participated in ten plenary meetings and twenty-three informal meetings at the conference. Sharp political divisions became apparent. Many African and Latin American states, as well as EU members, wanted the conference to adopt legally binding measures, including a prohibition on the sale of SALW to nonstate actors. Since African countries are the most severely affected by the illicit small arms trade, they were concerned about preventing the transfer of SALW to terrorists and insurgent groups. Several like-minded states—including Canada, Finland, the Netherlands, and Norway—and humanitarian NGOs, such as Amnesty International and Human Rights Watch, argued that the eradication of the illegal trade in SALW could not be accomplished without first establishing stronger regulations on the legal trade. They emphasized that states should accept responsibility for the uncontrolled proliferation of SALW, and should refrain from providing SALW to regimes with dubious human rights records.

Amnesty International and Human Rights Watch were particularly vocal in warning that the New York conference would produce woeful results if it focused solely on illicit transfers and if it did not derive binding agreements. These NGOs also complained that the human rights and humanitarian elements of the SALW problem had been left off the conference agenda, and that the conference had neglected to address the culpability of governments in supplying the SALW which has been used to commit war crimes. But most states at the conference were more interested in the problem of preventing the destabilizing accumulation of SALW, rather than in the humanitarian consequences of the SALW trade. This is a very divisive issue, because a situation which one state may perceive as a destabilizing accumulation of SALW, another state may view as a necessary acquisition of SALW for the purpose of national security.

In addition, the efforts of the humanitarian NGOs were countered by a minority of NGO activists representing the "firearms community" in seven countries. While most of the pro-gun NGOs preached the need for the responsible and safe ownership of firearms, the 4.5 million member NRA expressed its deep concerns that the conference would adopt a plan of action that would threaten the legitimate domestic rights of American citizens to own and use firearms.

In contrast to the extensive participation of NGOs in both the Ottawa Process banning APLs and the Rome Conference establishing the ICC, NGOs were excluded from important meetings and given limited time to present their viewpoints at the SALW conference. Lloyd Axworthy, the former Canadian Minister for Foreign Affairs, expressed his frustration with the marginalization of the humanitarian NGOs at the conference:

> As preparations for the UN conference progressed, it became obvious that there was a concerted effort by the UN bureaucracy and the diplomatic disarmament establishment to avoid adopting any of the methodology or lessons of the land-mine experience. This was apparently to be a traditional meeting of nation-states. NGOs would be on the margin. (Axworthy 2003, 348)

In his address to the conference, the US Under-Secretary of State for Arms Control, John R. Bolton, argued that the responsible use of firearms for hunting and sport shooting is an important element of American culture ("UN Conference" 2001). He emphasized that while the United States supports actions to stem the illicit trade in military-type SALW, it opposes any initiative that would challenge the Second Amendment of the US Constitution and limit the use of hunting rifles and pistols by American citizens. The US delegation mounted a fierce resistance to the proposal to establish a ban on the transfer of SALW to nonstate actors. Although the United States was widely viewed as standing alone on this issue, other governments which professed the legitimacy of armed struggle by subnational groups for the achievement of independence and self-determination also believed that states should be able to supply such groups with SALW.

Since the negotiations at the New York conference were conducted via consensus decision-making, the US delegation succeeded in eliminating references to the regulation of private gun ownership, as well as a ban on SALW transfers to nonstate actors, from the draft Program of Action (PoA). The United States did accept a requirement for governments to implement strict national laws and procedures on SALW, in order to minimize the diversion of SALW to illegal channels. But other major SALW exporters, such as China and Russia, joined the United States in rejecting any legally binding instrument that would enable the tracing of the lines of supply of SALW. Instead, these states approved weaker measures, including the strengthening of the capacity of states to cooperate in tracing illicit flows of SALW, and the launching of a UN study on the feasibility of developing an international instrument that would facilitate the tracing by states of illicit SALW transfers. The US objected, initially, to the setting of a sequence of follow-up activities to the conference, but finally agreed to biennial meetings of states and another global conference by 2006 in order to monitor progress.

On the final day of the conference, the participants adopted a PoA, which represents the first global commitment made by states to prevent and eliminate the illicit SALW trade. The PoA commits states to pass national laws criminalizing the illicit trade in SALW; regulate the activities of SALW brokers; require licensed SALW manufacturers to place traceable markings on each weapon produced; establish strict criteria for the export of SALW; prosecute violators; keep accurate records on the manufacture, possession, and transfer of SALW; and cooperate with other SALW initiatives at the global, regional, and national levels (NGO Committee on Disarmament 2001b). But the PoA is a voluntary agreement which is politically, but not legally, binding ("UN Conference" 2001). The like-minded states and humanitarian NGOs which had campaigned for more significant action on the SALW issue were deeply disappointed (NGO Committee on Disarmament 2001c). The African states were particularly upset with the failure to proclaim a ban on the transfer of SALW to terrorists and guerrillas.

But the need to reach a consensus agreement at the conference, in order to take some action on the illicit SALW trade, forced the like-minded states and humanitarian NGOs to accept a less ambitious PoA. In his final statement, the President of the Conference, Ambassador Camilo Reyes of Colombia, expressed his frustration with "the conference's inability to agree, due to the concerns of one state, on language recognizing the need to establish and maintain controls over private ownership of these deadly weapons and the need for preventing sales of such arms to non-state groups" (UN 2001, 23). Reyes also emphasized that the problem of the illicit trade in SALW must be addressed in all its aspects.

Jayantha Dhanapala (2002), the former UN Under-Secretary-General for Disarmament Affairs, emphasized that although the New York conference failed to address key issues related to the legal SALW trade, the conference should be measured by its contributions toward resolving the SALW proliferation crisis. First, the fact that a major UN conference was held on the issue of SALW a mere eight years after the issue was brought to the attention of the international community by Mali is remarkable. Dhanapala made the argument that "global norms are not built overnight" (Dhanapala 2002, 168). It should be noted that Dhanapala expressed his approval of the consensus decision-making of the New York conference, the benefits of which may have actually been outweighed by its costs, because the need to derive a consensus agreement allowed a small minority of detractor states the opportunity to prevent any significant action from being taken on the issue of the licit SALW trade.

Second, Dhanapala indicated that since the New York conference was the first UN conference on the SALW issue, it should not have been a surprise that most states were not yet ready to consider the issue of the legal trade in SALW. Tim Martin, the Director of the Peacebuilding and Human Security Division of the Canadian DFAIT, emphasized that "the presence of arms is more of a minus to human security than a positive to national security."[1] But a November 2001 conference organized in Nairobi by the Humanitarian Coalition on Small Arms concluded that the Rome Conference "came too early at a time when political will to seriously tackle the human cost of small arms proliferation and misuse is not fully developed. Clearly most states are not prepared to put human security before national security" (Humanitarian Coalition on Small Arms 2001). Dhanapala reassured that "the door, however, is still open for such an instrument and the subject will surely be discussed within the upcoming review meetings" (Dhanapala 2002, 168). Unfortunately, this prediction has not come true so far.

Third, Dhanapala argued that the PoA, even as a nonlegally binding agreement, does not condemn states to inaction on the issue of the licit SALW trade. Instead, the agreement encourages states, international organizations, and NGOs to exercise leadership and develop more substantive arrangements. The PoA may be viewed as merely the first in a sequence of steps toward the adoption of stricter measures on the SALW trade.

After the conference

In November 2001, the UN General Assembly adopted Resolution 56/24 V, *The illicit trade in small arms and light weapons in all its aspects*. The resolution welcomed the adoption of the PoA by consensus, expressed the support of the member states for action to halt the illicit trade in SALW, and called for the convening of a review conference in 2006, as well as biennial meetings beginning in 2003 (Dhanapala 2002; UN Department for Disarmament Affairs 2003). One year later, the General Assembly passed Resolution 57/72 on November 22, 2002, which stressed the importance of an early and complete implementation of the PoA, and called for the convening of the first biennial meeting in New York City in July 2003.

The United Nations First Biennial Meeting of States to Consider the Implementation of the UN Program of Action to Prevent, Combat and Eradicate the Illicit Trade in Small Arms and Light Weapons in All its Aspects at the National, Regional and Global Levels was held in New York City from July 7th to 11th, 2003. The purpose of the meeting was to discuss the successes and difficulties experienced by states, international organizations, and NGOs during the first two years of implementation of the PoA. But the meeting did not have a mandate to take action on the two issues which were not included in the PoA: controls over private ownership of SALW, and a ban on transfers of SALW to nonstate actors. Several delegations, including the African Group, Canada, the EU, IANSA, the Netherlands, and Norway, expressed their views on the PoA's shortcomings, and called on the international community to take stronger action on SALW (Canada 2003; EU 2003; IANSA 2003; The Netherlands 2003; Nigeria 2003; Norway 2003).

Delegations also voiced their concerns about aspects of the licit SALW trade at the second biennial meeting, which was held in New York City from July 11th to 15th, 2005 (Canada 2005; Nigeria 2005; Norway 2005; United Kingdom 2005; UNGA 2005b). But similar to the first UN meeting, the agenda was restricted to discussion of the global, regional, and national implementation of the 2001 PoA. The UN General Assembly did adopt an International Tracing Instrument on December 8, 2005. Although a nonbinding agreement, states have pledged to ensure proper marking and record-keeping of SALW, as well as cooperate in tracing illicit flows of SALW. Unfortunately, very few states have provided clear information as to their implementation of the instrument (Parker 2010).

From January 9th to 20th, 2006, a UN Preparatory Committee met in advance of the UN review conference on the implementation of the PoA. During the meetings, the Mexican delegation proposed that the review conference discuss a series of regulations on the civilian possession of SALW (UN Preparatory Committee 2006a). These restrictions included the issuing of licenses for the ownership and possession of SALW and ammunition solely to citizens who pass mandatory background checks and show proof of training in the safe usage and storage of SALW, and an outright ban on civilian ownership of any SALW which is designed for military usage (such as automatic and semiautomatic assault rifles, machine guns, and light weapons). In the meantime, the Brazilian delegation issued proposals regarding global controls on SALW transfers (UN Preparatory Committee 2006c). The Brazilians suggested that in addition to procedural-operative measures intended to enhance national control and oversight of SALW transfers, the review conference should address the need for normative guidelines for states to follow when authorizing the import and export of SALW. While the Canadian government did not bring up the issue of the legal SALW trade, it did suggest that the review conference adopt a comprehensive approach which addresses both the supply and the demand for illicit SALW (UN Preparatory Committee 2006b).

On June 19, 2006, one week before the review conference, the UK submitted a working paper which discussed the Transfer Controls Initiative (TCI) launched by the British government in 2003 (UN Conference 2006a). The intent of the TCI was to develop common international standards for the trade and transfer of SALW. Ten regional workshops on the TCI were held between April 2004 and November 2005, and more than 110 states had expressed support for the TCI. The British government urged the review conference participants to support the development of common guidelines regulating SALW transfers.

The review conference on the progress made in implementing the PoA was held in New York City from June 26th to July 7th, 2006 (UNGA 2006). During an informal meeting on June 29th, the conference participants decided that a nonpaper submitted two days earlier by the Conference President, Prasad Kariyawasam from Sri Lanka, would become the working paper of the conference. Among its many recommendations, the paper called for greater national regulation of the production, possession of, and trade in SALW (UN Conference 2006b). Unfortunately, the review conference ended with the participants unable to reach a consensus on an outcome document.

From July 14th to 18th, 2008, the third biennial meeting of states to discuss the PoA was held in New York City (UNGA 2008). Some states mentioned issues related to legal SALW which they perceived as important for the implementation of the PoA on illicit SALW, including controls on the production and supply of SALW, prohibitions on supplying SALW to nonstate actors and terrorists, and the matter of civilian possession of SALW. It was even proposed that the PoA be transformed into a legally binding accord. But no agreements on these issues were reached at the meeting.

National delegations returned to New York City for the fourth biennial meeting, held from June 14th to 18th, 2010. During the meeting, the Canadian government issued a brief statement that described the extent of Canada-US cooperation to combat the illicit trafficking of SALW. The statement contained a single sentence referring to the legal SALW trade: "we recognize the importance of the efficient and effective regulation of the legal trade in small arms and light weapons as recognition of the legitimate interests of industry, shooting sports hobbyists and collectors, and in order to prevent diversion to illicit markets" (Canada 2010, 1). Once again, some states brought up certain issues at the meeting which other participants would not agree to, including ensuring the responsible civilian possession of SALW, prohibiting the supply of SALW to nonstate actors and terrorists, and making the PoA a legally binding agreement (UNGA 2010).

Although no progress on the issue of regulating the legal SALW trade has been made at the PoA meetings, an alternative process appears to be more promising. In December 2006, the UN General Assembly adopted Resolution 61/89 calling for negotiations on a legally binding Arms Trade Treaty (ATT) (IANSA 2011; UNODA 2011; Varner 2009). Once adopted, the ATT will regulate the international transfer of all conventional weapons, including SALW and perhaps ammunition as well. The George W. Bush administration cast the sole vote against the resolution, expressing both its concern about possible loopholes in the ATT, as well as its preference for national controls on the arms trade rather than an international regulatory instrument. But the Barack Obama administration reversed course, and illustrated its support for the ATT by voting for the October 2009 General Assembly resolution which established a timetable for the ATT meetings. US support was made conditional, however, on the use of consensus decision-making during the ATT negotiations in order to prevent the creation of loopholes. The 2009 resolution subsequently mandated that the UN Conference on the ATT engage in decision-making by consensus.

The Open-Ended Working Group held meetings in New York City in 2009, and the Preparatory Committee followed suit in 2010 and 2011. The Preparatory Committee met again in February 2012 to discuss the procedures of the ATT conference, which will be held in July 2012. The UN member states will attempt to reach a consensus on a draft treaty at the conference, which they will then sign. Ratification of the ATT may take years though, as this process requires the approval of national legislatures.

Conclusion

The initiative to adopt restrictions on the legal trade in SALW has not been successful so far, despite the significant efforts of the like-minded middle powers and humanitarian NGOs. The problem is that the middle powers have relied on the consensus decision-making process of the 2001 UN conference and subsequent meetings on SALW. The United States mounted a fierce opposition to the initiative at the conference, because Washington perceived any attempt to restrict the licit SALW trade as a threat to a principal element of its national interest: the constitutional right of American citizens, under the Second Amendment, to bear arms. The need for the conference participants to reach a consensus on an accord made it easy for the United States to block the inclusion of regulations on the licit SALW trade. The final result was a nonlegally binding PoA that is restricted to the illegal aspects of the SALW trade. As Lloyd Axworthy argued, "a global treaty incorporating all the competing interests, working through the consensus-based UN system where big states have a virtual veto, is bound to be limited" (Axworthy 2003, 349).

The middle powers were willing to accept consensus decision-making, however, because they realized that any accord reached on the illicit SALW trade, no matter how flawed, was better than no deal. There were limits to which the middle powers could push restrictions on the legal trade without jeopardizing the conference negotiations on the PoA. This was because the legal SALW trade initiative never reached the "tipping point" that the APL and ICC initiatives did, when a strong majority of states endorsed the proposals and gave them powerful forward momentum. Many states view the legal trade in SALW as essential for national security, and several pro-firearms NGOs defend the rights of citizens to own, purchase, and sell small arms. In other words, the international community and the global NGO community are divided on the necessity of regulating the legal SALW trade.

There is hope that the problematic effects of the legal trade in SALW will be addressed in the near future. As Jayantha Dhanapala (2002) suggested, the conference on the illicit SALW trade may have been only the first step toward making real progress in curbing the global proliferation of SALW. The international community has recently commenced negotiations on an Arms Trade Treaty, which may soon lead to a legally binding commitment to regulate the licit trade in SALW. But it remains to be seen if a lowest common denominator ATT derived through consensus decision-making will be an effective instrument.

Note

1 Personal interview of Mr. Tim Martin, the Director of the Peacebuilding and Human Security Division of the Canadian DFAIT, Ottawa, Ontario, Canada, December 2, 2003.

CHAPTER SEVEN

The responsibility to protect

Somalia. Bosnia-Herzegovina. Rwanda. Kosovo. The mention of these places invokes memories of human suffering. Millions of people were killed, injured, or displaced during these post-Cold War conflicts. The reputation of the United Nations was tarnished as well for the organization's inability to prevent or halt the slaughter. The perpetrators of war crimes, genocide, ethnic cleansing, and crimes against humanity hid behind the principle of inviolable state sovereignty. While the sovereignty principle protected the nation-state from external interference and intervention, the security of certain human populations within these states was jeopardized as vulnerable groups became victimized by the state.

This chapter examines how the middle powers took the lead in redefining the rules of sovereignty, so that calamities such as Somalia, Bosnia, Rwanda, and Kosovo never occur again. The solution lay in the development of a new norm, termed the "responsibility to protect" (R2P and RtoP are common acronyms). The chapter is structured as follows. First, the Westphalian sovereignty principle and the concept of sovereignty as responsibility will be contrasted. Second, I will analyze the extraordinary contributions of two middle powers, Canada and Australia, to the establishment and proceedings of the International Commission on Intervention and State Sovereignty (ICISS). Third, the findings of the ICISS, presented in their 2001 report *The Responsibility to Protect*, will be explored. Fourth, I will examine how Canada played the role of R2P norm entrepreneur, and how the international community responded to the notion of R2P. Fifth, the reaction of the United States to the concept of R2P will be discussed. Finally, the chapter looks at the extent to which R2P has been applied with regards to contemporary conflicts, and investigates whether R2P has become a norm in international relations.

Westphalian sovereignty and sovereignty as responsibility

Stephen Krasner (1999) distinguished between four different meanings of sovereignty. "Domestic sovereignty" refers to how authority is organized within a state, and the degree to which those in authority exercise effective control. "Interdependence sovereignty" is the capacity of the domestic authority to regulate transborder movements. "International legal sovereignty" involves practices related to mutual recognition between states. The fourth conception of sovereignty is the most relevant one for our purposes. "Westphalian sovereignty" entails "political organization based on the exclusion of external actors from authority structures within a given territory" (Krasner 1999, 4). This principle of state sovereignty dates back to the 1648 Treaty of Westphalia, which ended the Thirty Years' War.

The Westphalian sovereignty principle was enshrined in the 1945 United Nations Charter (UN 2011a). Article 2(1) declares that all UN members have sovereign equality. Article 2(4) states that all UN members should refrain from threatening or using force against the territorial integrity or political independence of another state, or in any manner that is inconsistent with the purposes of the UN. Under Article 2(7), the UN does not have the authority to intervene in matters which lie within the domestic jurisdiction of any state and cannot force UN members to submit domestic matters to international mechanisms for the settlement of disputes, but the UN does retain the authority of applying enforcement measures to a state under Chapter VII. Thus, the corollary to the Westphalian sovereignty principle is the international norm of nonintervention in a state's domestic affairs, a norm which applies both to states and the United Nations as an international organization. The renowned former Australian foreign minister Gareth Evans made the blunt point that "sovereignty is a license to kill: what happens within state borders, however grotesque and morally indefensible, is nobody else's business (Evans 2009, 16).

In contrast to this conception of traditional sovereignty, which serves as the foundation of the United Nations system, lies the notion of "sovereignty as responsibility" (Bellamy 2009b; Cohen and Deng 1998; Deng et al. 1996). Alex Bellamy (2009b) illustrated that sovereignty as responsibility is based on two key propositions. First, individuals have inalienable human rights that are universal and have priority over the rights of national groups, a proposition which contradicts the concept of traditional sovereignty. Second, governments have the primary responsibility for protecting the rights of their citizens, but should they fail to protect, or should they abuse these rights themselves, then the international community acquires the rights and duties to take action to protect the citizens' rights.

While the idea of sovereignty as responsibility can be traced as far back as the 1776 US Declaration of Independence, Francis Deng has been credited for resuscitating this concept in the post-Cold War period (Bellamy 2009b). Deng, a former Sudanese diplomat who was appointed by UN Secretary-General Boutros Boutros-Ghali as his Special Representative on Internally Displaced People (IDPs) in 1993, wrote a book with Roberta Cohen which argued that the primary responsibility for protecting IDPs lies with the host state (Cohen and Deng 1998). If the host state is unable to fulfill its responsibility, it should request assistance from the international community, which then becomes responsible for assisting the IDPs. The international community may even intervene on behalf of the IDPs in extreme cases when the host state does not provide its consent. As Deng and his co-authors stated in another book, "sovereignty as responsibility means that national governments are duty bound to ensure minimum standards of security and social welfare for their citizens and be accountable both to the national body politic and the international community" (Deng et al. 1996, 211).

Canada, Australia, and the ICISS

The concept of sovereignty as responsibility was embraced by the International Commission on Intervention and State Sovereignty. The creation of the ICISS, as well as its effective operation, may be attributed to the considerable efforts of two middle powers, Canada and Australia. The Canadian government took initiative in opening communication on the contentious issue of sovereignty and humanitarian intervention (Axworthy 2003). Don Hubert and Jill Sinclair of the Canadian Department of Foreign Affairs and International Trade recommended that an international Commission, consisting of leading experts and politicians, should be created to conduct a one-year study of the tensions between sovereignty and intervention. Canadian Foreign Minister Lloyd Axworthy then brought up the idea with UN Secretary-General Kofi Annan, who was highly supportive, but was also skeptical about the Commission's prospects for success under UN sponsorship. Annan suggested that Canada should sponsor the Commission instead, while promising to endorse the initiative and to receive the Commission's final report in person.

While the Canadian government was able to provide funds to cover the administrative costs of the Commission, additional funding was needed to cover the costs of the Commissioners' travel and research (Axworthy 2003). Fortunately, several sources of funding appeared, including American foundations such as the John D. and Catherine T. MacArthur Foundation, the Carnegie Corporation of New York, the William and Flora Hewlett Foundation, the Rockefeller Foundation, and the Simons Foundation (ICISS 2001a). The governments of Switzerland and the United Kingdom were also generous in providing financial and other support to the Commission.

Appointing the Commissioners proved to be a more delicate task, as the Commission needed to be representative of different regions and views. Axworthy remarked that the Commission "could not appear captive to either pro-interventionist views or a Northern Hemisphere perspective" (Axworthy 2003, 191). The Canadian government invited two renowned figures to head the Commission: Gareth Evans, the President of the International Crisis Group in Brussels and the former Foreign Minister of Australia; and the Algerian diplomat Mohamed Sahnoun, who was Special Advisor to the UN Secretary-General, and had previously served as Special Envoy and Special Representative of the Secretary-General for conflicts in Ethiopia and Eritrea, the Great Lakes region of Africa, and Somalia (ICISS 2001a). The former Australian diplomat Ken Berry served as Executive Assistant to the Co-Chairs. With input from the Co-Chairs, the Canadian government appointed ten other distinguished Commissioners: Gisèle Côté-Harper and Michael Ignatieff from Canada, Lee Hamilton from the United States, Vladimir Lukin from Russia, Klaus Naumann from Germany, Cyril Ramaphosa from South Africa, Fidel V. Ramos from the Philippines, Cornelio Sommaruga from Switzerland, Eduardo Stein Barillas from Guatemala, and Ramesh Thakur from India.

In September 2000, the UN held its Millennium Assembly. Canadian Prime Minister Jean Chrétien took advantage of the high-profile occasion to announce that an independent International Commission on Intervention and State Sovereignty would be created, in response to Secretary-General Annan's call for a global consensus on how to respond in situations of severe violation of international humanitarian law (ICISS 2001a). On September 14th, Lloyd Axworthy launched the ICISS and mandated it with fostering a debate on the conundrum between intervention and state sovereignty, as well as crafting an international consensus on a proper response to humanitarian calamities, especially through the UN. The objective was for the ICISS to conclude its work within one year, so that the Canadian government could present the Commission's findings at the fifty-sixth session of the UN General Assembly in the fall of 2001.

On October 17, 2000, John Manley replaced Lloyd Axworthy as Canada's foreign minister. Manley established an international Advisory Board to provide political advice to the ICISS, as well as to assist in generating the necessary public support to accomplish the Commission's recommendations once they were released (ICISS 2001a). Chaired by the former Canadian foreign minister Axworthy, the Advisory Board consisted of fifteen distinguished individuals with reputable backgrounds in politics, academia, and fundraising. The Advisory Board met with the ICISS in London on June 22, 2001 for a lively and constructive debate.

The Canadian government also established a Secretariat for the ICISS (ICISS 2001a). Located at DFAIT in Ottawa, the Secretariat was headed by Executive Director Jill Sinclair and Deputy Director Heidi Hulan, who presided over a staff of six individuals. The Secretariat engaged in fundraising, organized the meetings of the ICISS and the roundtable consultations, directed the publication and circulation of the Commission's report as well as the various research papers generated by the global consultation process, and led diplomatic efforts to garner political support from national governments for the activities of the ICISS.

The ICISS Secretariat was aided in its activities by the International Development Research Centre (IDRC) in Ottawa, as well as the staff at Canadian embassies around the world (ICISS 2001a). In addition, an international research team was formed to assist the ICISS. The team was led jointly by the American scholar Thomas G. Weiss, co-director of the UN Intellectual History Project and professor at The Graduate Center of the City University of New York (CUNY), and the Zimbabwean lawyer Stanlake J. T. M. Samkange, a former speechwriter to UN Secretary-General Boutros Boutros-Ghali. While Samkange acted as rapporteur, aiding the ICISS with the draft report, Weiss and the Canadian policy advisor Don Hubert from the Peacebuilding and Human Security division of DFAIT were responsible for writing the various research essays presented in the supplementary volume to the ICISS report (ICISS 2001b). A Research Directorate was set up at the CUNY Graduate Center in New York City to assist the authors.

Between January and July 2001, the ICISS held eleven regional roundtables and national consultations around the world, in order to hear different perspectives on the issue of intervention and sovereignty and spark a debate (ICISS 2001a, 2001b). These forums were held in Ottawa on January 15th, Geneva on January 31st, London on February 3rd, Maputo on March 10th, Washington on May 2nd, Santiago, Chile on May 4th, Cairo on May 21st, Paris on May 23rd, New Delhi on June 10th, Beijing on June 14th, and St. Petersburg, Russia on July 16th. Interested NGOs and international organizations were invited to most of these consultations, though the meeting in Geneva had a separate roundtable for the UN and international organizations, while the consultation in Paris had a separate roundtable for French government officials and parliamentarians.

Each of the roundtables featured at least one of the Commission's Co-Chairs as well as a few of the Commissioners. Prior to each meeting, the ICISS distributed a paper outlining the Commission's perspective to the participants in order to initiate a discussion. The ICISS also invited beforehand specific participants to prepare papers and make presentations on issues related to intervention and state sovereignty. One participant at each consultation was asked to write a summary report of the meeting following its denouement. The papers produced at these roundtables proved to be an invaluable resource for the Commission in preparing its final report.

The Commissioners traveled to numerous capital cities to give briefings on a regular basis to interested governments. The ICISS also met with diplomatic missions in Ottawa and Geneva. On June 26–27, 2001, the Commission briefed Secretary-General Annan and members of the UN Secretariat, as well as representatives from several permanent missions.

The ICISS held five full meetings and one informal conference within the year (ICISS 2001a). The Commissioners met first in Ottawa on November 5–6, 2000. They reviewed a series of key questions, identified the main issues, and came to a decision on an effective approach. The Commissioners drafted an outline of a report and circulated it among themselves. Sometime after the Ottawa meeting, Co-Chair Gareth Evans derived the term "responsibility to protect" as a means of reconciling the competing objectives of protecting human rights and respecting state sovereignty (Bellamy 2010a). Evans also hoped that R2P would resolve four problems. First, the discourse had been overly focused on military intervention, to the neglect of other ways in which the international community could intervene. Second, the developing countries were concerned about the troubling prospects for interventionism to turn into neoimperialism and neocolonialism. Third, it was highly unlikely that new international legal instruments to govern intervention would either be agreed upon by the international community, or if adopted, be capable of protecting endangered populations. Finally, not enough attention had been placed on the notion that different actors have different responsibilities with regards to protecting populations.

An informal meeting of the Commission was then held in Geneva on February 1, 2001, where several Commissioners met in person and others connected via conference call in order to discuss the draft outline and the concept of R2P (Bellamy 2010a; ICISS 2001a). The participants viewed Co-Chair Evans' concept of R2P as a promising approach for addressing the four issues, although many felt that the military dimension of intervention was still receiving an excessive emphasis. The Commissioners continued their discussions on the outline at a full ICISS meeting in Maputo on March 11–12, 2001.

An expanded draft was produced in May 2001, which was distributed to the Commissioners for review and commentary. The draft was discussed further at a meeting in New Delhi on June 11–12, 2001. The Commissioners agreed on a significant revision to the substantive matters and structure of the report. A new draft was written and distributed in July 2001, which the Commissioners provided written comments on. The Co-Chairs Gareth Evans and Mohamed Sahnoun met for several days in Brussels in July 2001, in order to write a full-length draft based on the written feedback of the Commissioners. The Co-Chairs' draft report was circulated to the other Commissioners one week prior to the ICISS meeting in Wakefield, Quebec. From August 5–9, 2001, the Commissioners met in Wakefield to discuss the Co-Chairs' report, and ultimately endorsed it unanimously. A final full meeting of the ICISS was held in Brussels on September 30, 2001, in order to discuss the possible ramifications of the September 11th terrorist attacks for the report. Several modifications were then made to the report prior to publication.[1]

The Responsibility to Protect: the conclusions of the ICISS

In December 2001, the report of the ICISS titled *The Responsibility to Protect* (ICISS 2001a), and a supplementary volume which provided research on the issue of sovereignty and intervention, an extensive bibliography, and background information on the ICISS (ICISS 2001b), were both published. The report presented two core principles of the R2P norm. First, the notion of state sovereignty implied a responsibility on the part of the state to protect its population. Second, if a population suffers harm due to civil war, insurgency, repression, or the failure of a state, and the state is unable or unwilling to alleviate the situation, the R2P norm prevails over the norm of nonintervention.

The ICISS claimed that R2P entails three explicit responsibilities. First, there is the "responsibility to prevent." The primary causes of conflict need to be addressed in order to prevent any escalations to violence. The ICISS emphasized that prevention is the most critical element of R2P, and that all prevention options should be tried before any intervention is attempted. Second, R2P involves the "responsibility to react." Actors need to respond to situations of human suffering with appropriate measures. The ICISS insisted that "whenever possible, coercive measures short of military intervention ought first to be examined, including in particular various types of political, economic and military sanctions" (ICISS 2001a, 29). Third, there is the "responsibility to rebuild." Following a military intervention, actors should assist with the recovery, reconstruction, and reconciliation processes.

The ICISS report also discussed the contentious issue of military intervention for humanitarian purposes, which the Commission called "an exceptional and extraordinary measure" (ICISS 2001a, xii). The ICISS established a "just cause threshold" for humanitarian military interventions. Such interventions would only be justified in the event of, or imminent probability of, massive loss of life (through purposeful state action, state failure to prevent, or a failed state situation) or large scale ethnic cleansing (through killing, expulsion, terror, or rape).

Humanitarian military intervention would require that certain precautionary principles be followed. First, the intervention must be based on the right intention of halting or preventing human suffering, no matter if the intervening states have secondary motives as well. The ICISS stressed that multilateral operations, with support from regional organizations and the victimized population, would be best for assuring that the right intention is being served. Second, military intervention should only be used as a last resort, when all nonmilitary options have been evaluated and have been deemed insufficient for resolving the situation. Third, the military intervention must use proportional means, that is, its scale, duration, and intensity should be the minimum necessary to achieve the humanitarian objectives. Finally, there have to be reasonable prospects for success in securing the population, with the consequences of action not worse than the ramifications of inaction.

The ICISS also addressed the role of the United Nations in authorizing humanitarian military interventions. The Commission argued that the UN Security Council is the best organ for authorizing interventions, but needs to be improved in terms of its functioning. The P5 members were requested to refrain from using their veto power to block the authorization of humanitarian interventions which have support from the majority of UN members, and do not conflict with the vital national interests of the P5. Moreover, the ICISS stressed that if the Security Council should fail to authorize an intervention, or to deal with an emergency situation in a timely manner, then alternative options may be pursued. These options would include seeking authorization from the General Assembly instead, under the "Uniting for Peace" procedure, or permitting the appropriate regional organizations to intervene under Chapter VIII of the UN Charter, with subsequent authorization from the Security Council. Furthermore, the ICISS issued a warning that the credibility of the UN would become tarnished if the Security Council fails to act in the event of a grave threat to a population, and states intervene unilaterally without Security Council authorization to remedy the situation.

The Commission specified the operational principles for a successful humanitarian military intervention as well. First, an intervention needs to have specific objectives, a clear mandate, and sufficient resources. Second, the participants in the intervention must share a common military approach, have unity of command, and clear communications and chain of command. Third, the participants should adopt a limited and gradualist approach to the use of force, keeping in mind that the objective of the intervention is to protect a population rather than militarily defeat a state. Fourth, a successful intervention requires precise and proportional rules of engagement which correspond to international humanitarian law. Fifth, the intervening forces must accept that their self-protection cannot become the main objective of the mission. Finally, the intervening military forces need to coordinate their activities with those of humanitarian organizations.

In the conclusion of *The Responsibility to Protect*, the ICISS issued a set of recommendations for immediate action on R2P. The Commission recommended that the General Assembly adopt a draft declaratory resolution which expresses the fundamental principles of R2P and contains four elements. First, the resolution should emphasize the notion of sovereignty as responsibility. Second, it should stress that the R2P of the international community involves three separate responsibilities—to prevent, to react, and to rebuild—whenever a particular state proves to be unable or unwilling to fulfill its R2P. Third, the General Assembly resolution should define the threshold for determining whether claims of mass atrocities warrant a military intervention. That is, it should specify that cases involving large scale loss of life or ethnic cleansing, whether ongoing or halted, are sufficient justifications for the international community to intervene militarily. Fourth, the resolution needs to convey the precautionary principles that must be followed whenever military force is deployed for humanitarian objectives, namely "right intention, last resort, proportional means and reasonable prospects" (ICISS 2001a, 74).

The Commission also provided recommendations to both the Security Council and the Secretary-General. The Security Council was urged to adopt the principles for military intervention that were formulated in the ICISS report. The ICISS suggested that the Secretary-General consider the options at his disposal to implement the recommendations of the report, and consult with both the President of the Security Council and the President of the General Assembly to determine how to promote R2P in those two UN organs.

The international community responds to R2P

Canada as R2P norm entrepreneur

The ICISS was highly successful in generating a global debate on the recommendations in its report, as well as in galvanizing the world into action on the idea of a responsibility to protect. Cristina Badescu and Thomas Weiss remarked that:

> The developments since the release of the ICISS report in December 2001 suggest that R2P has moved from the prose and passion of an international commission toward being a mainstay of international public policy. It also has substantial potential to evolve further in customary international law and to contribute to ongoing conversations about the responsibilities of states as legitimate sovereigns. (Badescu and Weiss 2010, 356)

It took middlepowermanship, however, to get the ball rolling on R2P. Playing the role of norm entrepreneur (Finnemore and Sikkink 1998; Florini 1996), Canada adopted a two-track approach for fulfilling the R2P initiative (Bellamy 2009b). The first track involved using diplomatic persuasion to convince governments to endorse R2P through resolutions and declarations, in order to further the development of a R2P norm. The intention was to create a coalition of like-minded states (the "Group of Friends of the Responsibility to Protect," which Canada co-chaired together with Rwanda[2]) that would promote R2P through intergovernmental channels. The Canadian government organized numerous workshops with other governments and held meetings with the permanent missions at the UN. Through these consultations, Canada became convinced that a majority of states could be persuaded to endorse R2P, despite the vocal objections of a few states within the Non-Aligned Movement (NAM). The Canadian strategy for building a strong majority of supporters involved illustrating that R2P is more than solely military intervention, and includes critical tasks of prevention and rebuilding. Furthermore, states needed to be convinced that R2P involved significant restrictions on the use of force, to address the concern that military invasions would increase as states used humanitarian rhetoric to justify them. The Canadian government argued that higher thresholds for the use of force than the Security Council utilized in the 1990s should be adopted. In addition, while the ICISS report had suggested that the Security Council should be the *primary* authorizer of the use of force, Canada insisted that the Council should be the *exclusive* authorizer. This Canadian argument would become a key element of the 2005 World Summit's consensus on R2P.

The second track featured the mobilization of civil society in support of the R2P initiative. The Canadian government asked the World Federalist Movement-Institute for Global Policy (WFM-IGP) to hold roundtable discussions with NGOs, in order to collect feedback on the ICISS report and plan a course of action on R2P. These roundtables evolved into the "Responsibility to Protect—Engaging Civil Society" (R2P-CS) project, led jointly by WFM-IGP and Oxfam. The R2P-CS project took the lead in organizing consultations with NGOs worldwide, rallying civil society in favor of R2P, educating the public about R2P, and pressuring national governments to come onboard. Through the roundtable talks, the NGOs came to the conclusion that the most effective approach for convincing governments to adopt R2P would be to emphasize the elements of the ICISS report dealing with the prevention of conflict and the protection of civilians.

Alex Bellamy claimed that during the first couple of years following the release of the ICISS report, "Canada stood 'almost alone' in trying to persuade states to commit to R2P" (Bellamy 2009b, 70). When Canada presented a draft General Assembly resolution in 2002 that would have committed the Assembly to discussion of the ICISS report, the NAM blocked it. A second Canadian draft resolution calling on the UN Secretary-General to facilitate a debate on R2P was defeated as well.

But the momentum soon started to shift in Canada's favor. In 2003, UN Secretary-General Kofi Annan assembled a sixteen-member High Level Panel on Threats, Challenges and Change (HLP), and tasked it with evaluating contemporary threats to global peace and stability, assessing how well the UN has responded to those threats, and recommending ways of reforming the UN to provide collective security in the new millennium. Gareth Evans, who had been appointed to the HLP, was highly influential in convincing the HLP members to incorporate R2P (Bellamy 2010a).

Canada submitted a nonpaper to the HLP which emphasized that R2P should respect the sovereignty of states, and that the international community should only act when high thresholds for intervention have clearly been surpassed. The Canadian nonpaper differed from the ICISS report, in that the nonpaper stressed that authorization for intervention must come from the Security Council, and did not include either a code of conduct for the P5 or procedures for situations where the Council fails to act despite sufficient mass atrocities to justify intervention. Alex Bellamy referred to Canada's position as "R2P lite," in that the concept of R2P as described in the ICISS report was watered down in order to make R2P more appealing to the broader international community, and thus facilitate the forging of a consensus on R2P (Bellamy 2009b, 73).

One year later, the HLP issued its report titled *A More Secure World: Our Shared Responsibility*, which noted that in today's world, "it cannot be assumed that every State will always be able, or willing, to meet its responsibility to protect its own peoples and not to harm its neighbours" (UN 2004, 1). The report expressed clearly and strongly the HLP's view on the importance of R2P:

> We endorse the emerging norm that there is a collective international responsibility to protect, exercisable by the Security Council authorizing military intervention as a last resort, in the event of genocide and other large-scale killing, ethnic cleansing or serious violations of international humanitarian law which sovereign Governments have proved powerless or unwilling to prevent. (UN 2004, 66)

The recommendations of *A More Secure World* were endorsed by Secretary-General Annan in a foreword to the report (UN 2004). In March 2005, Annan issued his own report titled *In Larger Freedom: Towards Development, Security and Human Rights for All*, in which he called on the Security Council to adopt a resolution clarifying the principles to follow when authorizing the use of force, as well as to issue a pledge that the Council would abide by these principles (UNGA 2005a).

The World Summit

On the occasion of its fiftieth anniversary, the United Nations planned to hold a World Summit on September 14–16, 2005. Heads of state and government would gather at the UN Headquarters in New York City to reaffirm their faith in the United Nations as well as their commitment to the UN Millennium Declaration adopted five years earlier. Preparations for the summit began toward the end of 2004. Jean Ping, who served as President of the General Assembly and Foreign Minister of Gabon, requested that ten permanent representatives, including two from the Western middle powers of Australia and the Netherlands, serve as facilitators for the summit, with the assigned role of negotiating with the wider UN membership to reach consensus on points to be included in an Outcome Document of the summit (Bellamy 2010a).

The negotiations proved to be more contentious than either Jean Ping or Kofi Annan predicted (Bellamy 2009b). The proposal for a P5 code of conduct for casting Security Council vetoes was opposed by the P5 and most UN members, who viewed the veto as a check on Western interventionism. Thus, the code of conduct was quickly discarded. With regard to the issue of Security Council authorization, the United States and the United Kingdom argued that the Council has *primary authority*, implying that interventions outside of UN channels are possible. But most states adopted the stance of the Canadian government that the Security Council has *exclusive authority*, which would certainly curtail Western interventionism. The UN members also disagreed about the necessity of criteria for authorizing interventions. While several African states echoed the HLP and Kofi Annan in calling for criteria that would render Security Council decision-making more transparent, accountable, and legitimate, the suggestion of instituting criteria was opposed by the United States because it wanted to maintain its freedom to decide independently whether to intervene, and by China and Russia because of their concern that criteria may be misused to legitimize interventions that are not authorized by the Council.

The R2P skeptics were numerous, particularly in the developing world. India was at the forefront of those who argued that R2P was merely an excuse for the West to intervene in the domestic affairs of less developed countries, a belief that was particularly prevalent in Asia (Bellamy 2009b). The Group of 77 (G77) would only accept R2P if they perceived that the summit negotiations were making progress on the issue of development. Several UN members, including Algeria, Belarus, Cuba, Egypt, Iran, Pakistan, Russia, and Venezuela, tried to have all references to R2P dropped from the Draft Outcome (Stahn 2007). Their reasons for excluding R2P varied. Some states were concerned that the concept was overly vague and susceptible to abuse. Others claimed that R2P was incompatible with the UN Charter and international law. A few states felt that R2P should be expressed as a moral rather than a legal responsibility.

But the developing world did not speak with one voice. The G77 and NAM states were divided on R2P, as several members declared their support for the responsibility to protect, including Argentina, Benin, Chile, Congo, Ghana, Guatemala, Mexico, Peru, Rwanda, South Africa, Tanzania, and Zambia. The African Union had already endorsed a regional right to intervene through the "Ezulwini consensus" of March 2005, which had established the AU as the primary institution for peace and security in Africa, tasked with intervening to quell humanitarian crises even prior to seeking UN Security Council authorization, if necessary (Bellamy 2009b).

The incredible efforts of a few individuals who fought hard for R2P during the summit negotiations deserve mention. Allan Rock, the Canadian Permanent Representative to the United Nations, confronted R2P skeptics like Russia and India head on in order to prevent their individual objections from thwarting the agreements reached in the Outcome Document (Bellamy 2009b). With the blessing of Kofi Annan, Jean Ping and a group of UN officials worked throughout the negotiations on a parallel, and secret, final draft of the Outcome Document containing R2P, which they unveiled on September 12th. Annan believed that the permanent representatives to the UN would accept the final draft, despite some objections, rather than reject the document and notify their heads of state and government, who were about to arrive in New York City, that there was nothing for them to sign. In order to preempt any final obstructionist efforts from the US Permanent Representative to the United Nations, John Bolton, Ping informed US Secretary of State Condoleezza Rice directly about the final draft and gained her approval.

Additional credit for the inclusion of R2P in the Outcome Document must be given to John Dauth, the Australian Permanent Representative to the United Nations, who was assigned the role of summit facilitator by Jean Ping and received strong support during the negotiations from Alexander Downer, the Australian Minister for Foreign Affairs. Dauth dug in his heels and refused to budge on R2P in the face of last-minute dissent from some R2P skeptics. He was prepared to scuttle all other agreements in the Outcome Document at the eleventh hour if R2P was rejected (Bellamy 2010a).

Dauth's strategy proved to be successful. On September 15th, the General Assembly issued a draft resolution reiterating the various commitments made by the UN members in the World Summit Outcome Document (UNGA 2005c). Paragraphs 138 to 140 of the Outcome Document were dedicated to the responsibility to protect populations in the event of genocide, war crimes, ethnic cleansing, and crimes against humanity:

> 138. Each individual State has the responsibility to protect its populations from genocide, war crimes, ethnic cleansing and crimes against humanity. This responsibility entails the prevention of such crimes, including their incitement, through appropriate and necessary means. We accept that responsibility and will act in accordance with it. The international community should, as appropriate, encourage and help States to exercise this responsibility and support the United Nations in establishing an early warning capability.
>
> 139. The international community, through the United Nations, also has the responsibility to use appropriate diplomatic, humanitarian and other peaceful means, in accordance with Chapters VI and VIII of the Charter, to help protect populations from genocide, war crimes, ethnic cleansing and crimes against humanity. In this context, we are prepared to take collective action, in a timely and decisive manner, through the Security Council, in accordance with the Charter, including Chapter VII, on a case-by-case basis and in cooperation with relevant regional organizations as appropriate, should peaceful means be inadequate and national

> authorities manifestly fail to protect their populations from genocide, war crimes, ethnic cleansing and crimes against humanity. We stress the need for the General Assembly to continue consideration of the responsibility to protect populations from genocide, war crimes, ethnic cleansing and crimes against humanity and its implications, bearing in mind the principles of the Charter and international law. We also intend to commit ourselves, as necessary and appropriate, to helping States build capacity to protect their populations from genocide, war crimes, ethnic cleansing and crimes against humanity and to assisting those which are under stress before crises and conflicts break out.
>
> 140. We fully support the mission of the Special Adviser of the Secretary-General on the Prevention of Genocide. (UNGA 2005c, 31–2)

The General Assembly adopted Resolution 60/1 with the consensus of 174 states on October 24, 2005 (Bassiouni 2009). Although the General Assembly endorsed the notion of R2P, it did not adopt all of the recommendations of the ICISS, including the establishment of criteria to guide interventions, the initiation of a code of conduct for the use of the veto by the P5 in the Security Council, and the exploration of the possibility of launching interventions without Security Council authorization (Bellamy 2010a). But the World Summit did define the scope of R2P. The UN members accepted Pakistan's arguments that R2P should be applied solely to four sets of crimes—genocide, war crimes, ethnic cleansing, and crimes against humanity—that were clearly defined under international law. The World Summit also established the rights and responsibilities of individual states (which have the primary R2P for their own populations) and the international community (which has the secondary R2P, should an individual state fail to fulfill its own R2P). Moreover, these rights and responsibilities are perennial.

The Security Council and R2P

Shortly after the World Summit, Secretary-General Annan recommended that the Security Council endorse R2P (Bellamy 2009a). The Council proceeded to initiate a contentious debate on the issue of R2P, which endured for six months. On December 1, 2005, the United Kingdom drafted a resolution which stressed the importance of the R2P provisions adopted at the World Summit. In addition, the British draft resolution emphasized that peace operations should protect civilians and allow the safe provision of humanitarian assistance, and also urged states to ratify human rights instruments, bring war criminals to justice, and protect refugees and internally displaced people.

The British draft resolution met considerable resistance. Security Council members Algeria, Brazil, China, the Philippines, and Russia backed away from their commitments to R2P pledged at the World Summit, and argued that the summit had only bound the General Assembly, and not the Security

Council, to engage in further discussion of R2P. Egypt (a nonmember of the Security Council) remarked that the General Assembly deliberations on R2P had not even commenced yet. Both Egypt and India (another nonmember of the Council) suggested that the debate on R2P should be transferred back to the General Assembly, and Russia supported their position, claiming that any Security Council pronouncements on R2P would be premature. While China declared publicly that it would be willing to endorse R2P as long as the Security Council would be the organ making the decisions to intervene under this norm, the Chinese government privately backed the position that further discussions on R2P should continue in the General Assembly (Bellamy 2009a).

But the R2P skeptics were countered by numerous R2P proponents, including Canada, the EU, Japan, South Korea, and a few developing countries such as Benin, Nepal, Peru, Rwanda, and Tanzania. Fortunately, the January 2006 turnover in the Security Council membership favored the R2P supporters. Algeria and Brazil, who were publicly hostile to the notion of R2P, left the Security Council, as did the Philippines, who usually sided with the anti-R2P states. Each of the five incoming members—Congo, Ghana, Peru, Qatar, and Slovakia—had expressed their support for R2P.

Most of the permanent members of the Security Council continued to have reservations, however. Russia was concerned that the Council would adopt R2P prematurely, and the United States was apprehensive that any endorsement of R2P would commit the Council to authorize more cases of intervention. But China began to modify its position on R2P, and expressed that it would support the Security Council's endorsement of R2P, but only if the Council's resolution mirrored the October 2005 General Assembly resolution. The United Kingdom took advantage of the shift in the Chinese stance to negotiate a new draft resolution, which was unveiled on April 13, 2006. The British complied with the interests of China by reducing the draft resolution to a mere endorsement of the salient points adopted by the General Assembly in the World Summit Outcome Document. On April 28, 2006, the Security Council adopted Resolution 1674 unanimously, thereby proclaiming its public support for R2P and the General Assembly resolution of October 2005 (Bellamy 2009a).

The United States' reaction to R2P

According to Lee Feinstein and Erica De Bruin (2009), the US government's perspectives on the issues of R2P and humanitarian intervention have developed considerably throughout the post-Cold War period. As a presidential candidate, Bill Clinton was critical of the American response to the mass atrocities in Bosnia, and was determined to provide US support to the UN in its efforts to counter genocide. But once he came into office, Clinton turned his attention to domestic issues. He was hesitant to intervene in Bosnia without the support of European allies and US military leaders, as well as the favorable public opinion of American citizens. The Clinton administration also withdrew four thousand US troops from Somalia following the deaths of eighteen American Rangers in October 1993. In addition, the Clinton administration was reluctant to intervene in Rwanda in 1994, and refrained initially from describing the situation there as "genocide."

Despite dropping the ball with regards to the three worst cases of humanitarian crises in the 1990s, the Clinton administration did take measures to improve the United Nations' capacity for peace support operations. These included paying around $400 million of the American arrears in peacekeeping dues, assisting with upgrades at the UN's Department of Peacekeeping Operations, advicing the UN on how to enhance its logistics and communications capacities for peace support missions, and establishing an American program to train African soldiers as UN peacekeepers and provide them with proper equipment.

But while it was committed to reforming the UN, the Clinton administration demonstrated that it was willing to act outside UN channels when it led the NATO intervention in Kosovo in 1999, to protect Kosovar Albanians from mass murder and ethnic cleansing by the Serbian military. Widely criticized for this infringement of Yugoslavia's sovereignty without Security Council authorization, Clinton defended his actions by stressing the need to save the lives of innocent civilians, and emphasized the risks that the Kosovo conflict would spark violence throughout the Balkans. Although Clinton pledged that the United States would stop future genocides anywhere in the world, he stopped short of adopting formal guidelines for humanitarian intervention, preferring to deal with potential situations on an ad hoc basis.

The George W. Bush administration was hostile to the idea of humanitarian intervention, even in situations of mass atrocities. Bush was adamant about avoiding any agreements that would commit the United States to intervene in countries that were deemed peripheral to American national interests. Furthermore, the attacks of September 11, 2001, and the launch of the "War on Terror" absorbed the full attention of Washington, thereby relegating the issue of humanitarian intervention to the back burner. The Bush administration was disinterested in any new international institutions that could potentially interfere with how the United States waged the War in Afghanistan and sought reprisals against Al-Qaeda. Thus, the Bush administration expressed its displeasure with the ICISS report in May 2002, and rejected the principle of R2P (Feinstein and De Bruin 2009).

But the Bush administration's policy on R2P would soon change drastically. R2P and the War on Terror shared a common preoccupation with the threats posed by both predatory regimes and failed states. The Bush administration quickly picked up the same ideas of prevention, reaction/intervention, and rebuilding which had been expressed in the ICISS report, and applied them as part of the American global antiterror campaign.

Moreover, the growing concern of American citizens with the humanitarian crisis in the Darfur region of Sudan put pressure on the Bush administration to act. The administration was hesitant to intervene due to concerns about jeopardizing the United States' relations with the Sudanese government, which was viewed as a critical partner in the War on Terror. Evangelical Christians, ardent supporters of the Republican Party, called on the Bush administration to intervene in Darfur, militarily if necessary, to stop the massacres. In July 2004, Congress declared the conflict in Darfur to be genocidal, and appealed to the president to follow suit in recognizing the genocide, as well as to place sanctions on the Sudanese regime (Feinstein and De Bruin 2009).

The following year, a task force led by former House Speaker Newt Gingrich (R-Georgia) and former Senate Majority Leader George Mitchell (D-Maine) issued a report endorsing R2P (Gingrich and Mitchell 2005). The Gingrich-Mitchell report emphasized that the primary R2P commits states to protect populations living on their territory from genocide and mass atrocities, as well as to not engage in genocide or mass atrocities themselves. The international community has the secondary R2P, and may act in the event that a state does not fulfill its primary R2P. In addition, the Gingrich-Mitchell report stressed that the international community should be willing to intervene to fulfill its R2P even if the Security Council fails to act. Although the reaction of the Bush administration to the Gingrich-Mitchell report was supportive, it remained silent with regards to the notion of R2P (Feinstein and De Bruin 2009).

As a result, the international community was uncertain about the US position on R2P prior to the 2005 World Summit. Moreover, the Bush administration employed an inconsistent approach in its relations with the UN. The administration had shocked the world in 2003 by launching an invasion of Iraq, without the authorization of the Security Council, in order to oust the Saddam Hussein regime. Although the Bush administration had shown its disdain for the responsibilities and restrictions of multilateralism, it was actively campaigning for the reform of the UN, the preeminent multilateral institution in the world. But while the administration demonstrated that it was interested in UN reform, it made the contradictory decision of appointing the staunchly anti-UN John Bolton as the US Permanent Representative to the United Nations one month before the World Summit.

The United States played an active role during the summit negotiations on R2P, with two objectives (Feinstein and De Bruin 2009). First, the United States wanted to guarantee that any agreement on R2P would introduce a moral responsibility to protect, but not a legally binding one. The Bush administration drew from the Gingrich-Mitchell report in emphasizing that the R2P of the international community is secondary to the R2P of individual states. Second, the Bush administration wished to maintain the freedom of individual states to decide on an ad hoc basis how to respond to humanitarian crises whenever they arise. The United States opposed the idea that states should be committed to automatic interventions whenever certain criteria are reached, as the ICISS, Secretary-General Kofi Annan, Canada, the United Kingdom and other pro-R2P states had proposed. In addition, the Bush administration argued that states may occasionally need to intervene without Security Council authorization, such as in a situation where a humanitarian crisis spills over national borders and provokes another state to act in self-defense.

But Ambassador John Bolton frequently played hardball during the World Summit negotiations, which only served to fuel the R2P skeptics and disrupt consensus-building on R2P. On August 30, 2005, Bolton sent a letter to the UN member states, which called for certain revisions to the R2P section of the draft Outcome Document that would make it compatible with the US position. Bolton insisted that the R2P of other countries is not equivalent to the R2P of the state where the humanitarian crisis is occurring, that R2P should be defined as a moral rather than a legal responsibility, that the Security Council does not have the legal obligation to protect populations that host states do, that references to criteria for intervention should be dropped, and that the possibility of intervention without Security Council authorization should be ensured (Bellamy 2009b; Stahn 2007).

The pro-R2P bloc of states complained, however, that the American proposals would be overly restrictive, and insisted that any last minute changes to the draft Outcome Document would potentially thwart the agreements which had already been negotiated (Feinstein and De Bruin 2009). In the end, Bolton's opposition to the Outcome Document was overruled by US Secretary of State Condoleezza Rice (Bellamy 2009b). On September 15th, the draft resolution of the World Summit Outcome was referred to the High-level Plenary Meeting of the General Assembly, and the Assembly adopted it through consensus as Resolution 60/1 on October 24th. The United States joined 173 other states in voting for the resolution (Bassiouni 2009). Following the World Summit, Ambassador Bolton addressed the US Senate Committee on Foreign Relations and described how the United States influenced the R2P negotiations:

> We were successful in making certain that language in the Outcome Document guaranteed a central role for the Security Council. We were pleased that the Outcome Document underscored the readiness of the Council to act in the face of such atrocities, and rejected categorically the argument that any principle of non-intervention precludes the Council from taking such action. (American Society of International Law 2006, 464 [hereafter ASIL])

While Paragraph 139 of the World Summit Outcome Document does state that the international community is "prepared to take collective action, in a timely and decisive manner, through the Security Council" (UNGA 2005c, 31), Lee Feinstein and Erica De Bruin (2009) claimed that since there is no explicit mention of Security Council or General Assembly authorization being required, there is a possibility that individual states or coalitions of states may exercise their R2P outside of UN channels, either unilaterally or through regional organizations. Carsten Stahn (2007) indicated that the responsibility to take collective action via the Security Council is dependent both on states participating in such action on a voluntary basis, and on states evaluating whether a particular case warrants a Security Council decision regarding R2P. Thus, the states which fought against the adoption of a legal R2P achieved their objective. Furthermore, the Outcome Document does not include any specific criteria for the use of force in the event of a humanitarian crisis. The Bush administration was certainly pleased with the inclusion of these points that corresponded with American interests.

According to Feinstein and De Bruin (2009), although the United States was not among the leading states which advocated for the adoption of R2P, American support for R2P was crucial for the success of the initiative. The Bush administration's insistence that the primary responsibility to protect lies with individual states rather than with the international community was a particularly strong argument that convinced holdout states to throw their support behind the R2P initiative. The World Summit Outcome Document ultimately incorporated this American perspective. Most of the developing world was concerned about the possibility that the great powers would use arguments about R2P to justify neoimperialist interventions in their domestic affairs. This was especially worrisome in the context of the Iraq War, considering that the Bush administration had used dubious humanitarian rhetoric to support its intervention to oust the Saddam Hussein regime (Roth 2009). But the fact that the United States itself expressed concerns publicly about R2P eased fears among less developed countries that R2P was merely a new justification for great power infringements of their sovereignty.

US support for R2P has continued during the Barack Obama administration. When he was a presidential candidate in 2007, Obama wrote an article in *Foreign Affairs* in which he stressed the need to use American military force to halt mass atrocities, as well as the importance of receiving support and participation from other states in these humanitarian efforts (Obama 2007). The Obama administration has emphasized the centrality of the United Nations and multilateralism to American foreign policy, and with regards to R2P, has assembled "a national security team [that] promises to turn rhetoric into action" (Weiss 2009, 150). In her address to the UN Security Council on January 29, 2009, the newly appointed US Permanent Representative to the United Nations, Susan Rice, reiterated the United States' commitment to the R2P tasks of prevention, reaction, and rebuilding (International Coalition for the Responsibility to Protect 2009 [hereafter ICRtoP]). Both the Quadrennial Defense Review and the National Security Strategy of 2010 emphasized the importance of being prepared to act in the event of mass atrocities or genocide (US Congress 2010). On December 22, 2010, the US Congress adopted a concurrent resolution which recognized that preventing genocide and other mass atrocities is in the American national interest, and called for a "whole of government approach" to prevent and halt these crimes (US Congress 2010).

R2P in action

In August 2007, Secretary-General Ban Ki-moon announced that he would appoint Edward Luck as Special Adviser on the R2P (Bellamy 2009b, 2010b). But the General Assembly would only confirm Luck's position with the removal of the phrase "R2P" from his job title, due to strange arguments that the Assembly had only consented during the World Summit to further *discussions* on R2P, not to the principle of R2P itself. After assuming his position in 2008, Luck was given the tasks of deriving recommendations for operationalizing R2P within the UN system, and of consolidating the international consensus on R2P. Despite limited resources, Luck did an effective job in consulting with national governments on R2P, and his work paved the way for the Secretary-General to develop a "narrow but deep" approach for promoting R2P.

On January 12, 2009, Secretary-General Ban issued his report titled *Implementing the Responsibility to Protect* (UNGA 2009). The report outlined a three-pillar strategy for advancing R2P, where Pillar One emphasized the state's R2P regarding its own population, Pillar Two referred to the international community's commitment to assisting states in fulfilling their R2P, and Pillar Three stressed the responsibility of the international community to intervene whenever a state fails to protect or threatens its population. The General Assembly then held a debate on R2P in September 2009, which featured ninety-four speakers representing around 180 governments (including those belonging to the NAM). Cuba, Nicaragua, Sudan, and Venezuela joined the President of the General Assembly, Father Miguel d'Escoto Brockmann from Nicaragua, in calling for the rejection of R2P. But the other UN members concurred that the objective should be to implement R2P, and not to renegotiate the agreements reached at the World Summit (Bellamy 2010b). The General Assembly passed a resolution which endorsed Secretary-General Ban's report, and promised to continue deliberations on R2P. On November 11, 2009, the Security Council also reiterated its commitment to R2P through Resolution 1894.

But ever since the Security Council adopted Resolution 1674 unanimously on April 28, 2006, illustrating its support for R2P for the first time, only three cases concerning R2P have been brought up in the Council. The first case was in August 2006, when the Security Council passed Resolution 1706, which mentioned R2P in the case of the mass atrocities in the Darfur region of Sudan (Bellamy 2010b). There was no dissension in the Security Council over this resolution, and the outcome was the authorization of the African Union/United Nations Hybrid operation in Darfur (UNAMID), as well as the referral of the situation to the International Criminal Court.

The second case occurred on February 26, 2011, when the Security Council adopted Resolution 1970 concerning the conflict in Libya. The resolution mentioned the responsibility of the Libyan government to protect its citizens, implemented an arms embargo, asset freeze, and travel ban on the

Muammar Gaddafi regime, and referred the situation to the ICC (UN Security Council 2011a). The situation in Libya provoked the UN Human Rights Council to refer to R2P for the first time in Resolution S-15/1, which led the General Assembly to pass Resolution 65/60, suspending Libyan membership in the Human Rights Council (Weiss 2011). The third case was Security Council Resolution 1996 of July 8, 2011, welcoming the independence of South Sudan. The resolution established the UN Mission in the Republic of South Sudan (UNMISS), which was tasked, among other objectives, with assisting the South Sudanese government in fulfilling its responsibility to protect civilians (UN Security Council 2011b).

Although the Security Council has been reluctant to refer to the responsibility to protect, R2P has been invoked in several humanitarian crises since the 2005 World Summit by other actors (Bellamy 2010b). These cases include the postelection violence in Kenya in 2007 and 2008, Georgia's military actions in South Ossetia in 2008, Myanmar's failure to respond to the humanitarian tragedy in the wake of the May 2008 Cyclone Nargis, the situation of the Palestinians in the Gaza Strip in 2009, the crackdown on the Tamils in Sri Lanka in 2008 and 2009, the ongoing situation in the Democratic Republic of the Congo, the totalitarian repression of the North Korean regime, and the Myanmar military government's oppression of ethnic minorities. But while R2P was brought up by either UN officials or national governments, discussions of these situations within the UN organs were few or none, and no Security Council authorization was given for any interventions. Moreover, a couple of the accusations of R2P crimes were widely rejected, namely the baseless Russian claims of Georgian atrocities in South Ossetia, and the French attempt to link the aftermath of the natural disaster of Cyclone Nargis to a failure by the Myanmar regime to protect its population.[3]

There have also been some humanitarian crises where there was clear evidence that R2P crimes were committed, but R2P was never invoked (Bellamy 2010b). The conflict between Northern and Southern Sudan killed more civilians during the period from 2008 to 2009 than the Darfur conflict, and displaced more than 250,000 people from their homes. Since 2006, the conflict in Somalia has killed around 16,500 civilians, and has displaced approximately 1.9 million people. The wars in Afghanistan and Iraq also featured high civilian casualty rates and mass atrocities. But the international community failed to exercise its R2P in these cases where the state has either been the catalyst of the humanitarian crisis, or has been unable or unwilling to halt the atrocities.

The fact that actions in the name of the responsibility to protect have been hesitant and inconsistent ever since the General Assembly and the Security Council both endorsed R2P leads to the question if R2P has actually become a norm in international relations? Thomas Weiss, a prominent R2P scholar, claimed that "the responsibility to protect qualifies as emerging customary law after centuries of more or less passive and mindless acceptance of the proposition that state sovereignty was a license to kill" (Weiss 2009, 149). Alex Bellamy (2009b), another leading expert on R2P, indicated that different actors have labeled R2P as a concept, a principle, or a norm. Most governments (whether supportive or critical of R2P), as well as Edward Luck (the Secretary-General's Special Adviser on matters relating to R2P), have described R2P as a concept. In their view, R2P is an abstract idea that requires further elaboration before it can develop into a norm. The ICISS and the HLP have both referred to R2P as a principle, defined by Stephen Krasner as a "[belief] of fact, causation, and rectitude" (Krasner 1983, 2). Scholars have preferred to call R2P a norm, in the belief that a common understanding has emerged in the international community as to the appropriate behavior of states with regards to R2P.

But Bellamy himself has been inconsistent in his characterization of R2P. In his 2009 book *Responsibility to Protect*, Bellamy described R2P as both a concept and a principle, but not a norm, in order to correspond with "the terms in which the governments themselves refer to the R2P" (Bellamy 2009b, 7). He argued that R2P was a concept when it was formulated by the ICISS, and became a principle after it was modified during the World Summit negotiations and then adopted by the General Assembly. However, in a 2010 article in the journal *Ethics & International Affairs*, Bellamy claimed that "there is general consensus that RtoP is a norm, but much less agreement on what sort of norm it is" (Bellamy 2010b, 160). Moreover, Bellamy added that "R2P is not a single norm but a collection of shared expectations that have different qualities" (Bellamy 2010b, 160). The three pillars of R2P, as described by Secretary-General Ban Ki-moon, need to be analyzed separately. The first pillar, that states should refrain from carrying out genocide, war crimes, ethnic cleansing, and crimes against humanity, has been an embedded norm of international relations for decades. But the second pillar, that states should assist each other in fulfilling their R2P, and the third pillar, that the international community should act in the event that a state does not follow through on its R2P, are indeterminate and do not qualify as norms because states insincerely "adopt the norm's language while persisting with established patterns of behavior" (Bellamy 2010b, 166). Bellamy concluded that it would be better to perceive R2P as a policy agenda that needs to be implemented, rather than as a motivation for the international community to intervene altruistically.

Conclusion

Middlepowermanship has clearly been a catalyst in the evolution of sovereignty in the post-Cold War period. Middle powers played the roles of norm entrepreneurs in launching the R2P initiative and building global support for it. The concept of a responsibility to protect was derived by the former

Australian foreign minister Gareth Evans. Canadians and Australians made monumental contributions to launching the ICISS, conducting global consultations on R2P, and producing the landmark ICISS report. Canadian and Australian diplomacy were highly effective in organizing the pro-R2P states as the Group of Friends of the Responsibility to Protect, cultivating a close relationship with civil society groups in order to promote R2P worldwide, and ensuring that the R2P skeptics did not destroy the strong consensus on R2P that emerged from the World Summit negotiations (i.e. the "norm cascade," using Finnemore and Sikkink's [1998] terminology). It is evident that middle power leadership transformed R2P from a concept to a principle. R2P is currently in Finnemore and Sikkink's (1998) third stage of norm evolution. Whether or not R2P will develop into a norm of international relations depends on the degree to which states "internalize" R2P, and adapt their behavior to respect it.

While the United States was initially hostile to the idea of R2P, Washington eventually embraced the concept when it realized that R2P was consistent with the American national interest of preventing genocide and other mass atrocities around the world. The United States did negotiate with a heavy hand during the World Summit discussions in order to ensure that the R2P principle that emerged was significantly modified from the original concept of R2P as expressed in the ICISS report. But ultimately the United States did acquiesce to the R2P principle, and has committed itself to addressing R2P crimes globally.

The middle powers relied on the consensus decision-making process of the summit negotiations so as to generate as widespread an acceptance of R2P as possible, in the hope that inclusiveness would help build a solid R2P norm. The price of consensus decision-making was that the strong conception of R2P as presented by the ICISS was watered down (including by Canada) to become acceptable to the vast majority of participants at the summit. Nevertheless, a consensus did emerge on the idea of R2P, and the international community adopted R2P as a principle to guide interventions to halt genocide and mass atrocities. These incredible achievements would have never been accomplished without the leadership of the middle power states.

Notes

1 Alex Bellamy (2010a) indicated that the final report of the ICISS was written by Co-Chair Gareth Evans and Commissioners Ramesh Thakur and Michael Ignatieff.

2 The Netherlands recently replaced Canada as co-chair of the Group of Friends of the Responsibility to Protect. See Advisory Council on International Affairs 2010.

3 According to Bellamy (2010b), the international community reached a consensus that R2P is not applicable in the event of a natural disaster.

CHAPTER EIGHT

Conclusion

This book has illustrated that it is possible for middle power states to exercise effective leadership on global security issues. Middle power leadership has achieved human security objectives in the post-Cold War era, even when faced with great power opposition, which contradicts the arguments made by realists that "lesser states" should automatically heed the dictates of the great powers when it comes to matters of international security. Moreover, successful middle power leadership on the human security agenda challenges the erroneous belief, held by some scholars of middlepowermanship, that middle powers are restricted to playing supportive "follower" roles in the domain of security. In contrast, the middle powers have played the important roles of "securitizing actors" who focus global attention on how human populations are threatened existentially by certain phenomena, and of "norm entrepreneurs" who initiate campaigns to promote the adoption of human security norms by the international community.

I begin this chapter with a review of the five human security initiatives. The middle powers' choices of diplomatic strategy are discussed, as are the initiatives' impacts on the US national interest and the American reaction to each initiative. I also examine how successful, or not, these initiatives have been. Finally, I conclude by evaluating the contemporary status of human security, both as a theoretical approach and as a motivation for foreign policy action.

Assessing the human security initiatives

The middle powers have a few major accomplishments under their belt in the realm of human security (see Table 8.1). Middle power states addressed the need for a UN rapid response capability for peacekeeping missions by creating SHIRBRIG. The Ottawa Process, which produced a total ban on the use, production, stockpiling, and trade in APLs, was driven by middlepowermanship.

Table 8.1 The human security initiatives

Initiative	Middle power strategy	Impact on US national interest	US position	Outcome
SHIRBRIG	Fulfilled initiative without great power participation	Supported US interest in UN peacekeeping reform	None expressed	SHIRBRIG created in 1996 and deployed, but terminated in 2009
APL ban	Used fast-track diplomacy during Ottawa Process	Challenged US security interests in Korea	Acquiesced to ban, though nonparty to Ottawa Convention	APL ban took effect in 1999
ICC	Used soft power to control Rome Conference	Perceived as challenging US constitutional rights and threatening US military personnel abroad	Opposed to Rome Statute, but positive assessment of ICC in action	ICC came into effect in 2002
Regulating legal SALW trade	Relied on consensus decision-making	Perceived as challenging US constitutional rights	Opposed	Unsuccessful so far, but negotiations underway on the ATT
R2P	Relied on consensus decision-making	Supported US interest in genocide prevention	Acquiesced	Adopted at 2005 World Summit, but modified from original ICISS concept

Middle power leadership was instrumental for the establishment of an independent ICC which is empowered to carry out investigations and prosecutions of crimes against humanity, war crimes, genocide, and the crime of aggression. The middle powers also proposed a redefinition of sovereignty that emphasizes the responsibility of states to protect their citizens, which the international community endorsed enthusiastically. The sole initiative which has so far failed to achieve its objectives is the attempt to derive stricter regulations on the legal trade in SALW. But a record of four middle power victories in five attempts is nothing to scoff at.

The strategies of the middle powers

The middle powers have employed a variety of strategies in order to fulfill their human security initiatives. The establishment of SHIRBRIG was a one hundred percent middle power initiative, in that it was accomplished without the participation or assistance of any great powers. This was a deliberate move, as there were fears that great power involvement would lead to great power cooptation and manipulation of the brigade. Instead, middlepowermanship created a rapid reaction force for UN peacekeeping that could be deployed independently of the interests of the great powers. Canada and the Netherlands set up a coalition of like-minded states—the Friends of Rapid Deployment—which promoted a Danish proposal for a standby brigade that would be available, on short notice, for UN traditional peacekeeping operations. More than a dozen middle powers proceeded to pool their resources to establish SHIRBRIG.

In order to achieve the APL ban, the middle powers used fast-track diplomacy. Only those states and NGOs which supported a legally binding APL ban were invited to participate in the October 1996 conference and sign the Ottawa Declaration. Canadian Foreign Minister Lloyd Axworthy then challenged the international community to negotiate and sign an APL ban treaty within a mere fourteen months. The Ottawa Process core group organized a massive pro-ban coalition of like-minded states, international humanitarian organizations, and NGOs, which used both intergovernmental diplomatic channels and grassroots activism to campaign for a total ban on APLs, and to resist attempts by some states to water down the treaty through the inclusion of exemptions. Their efforts paid off as the Ottawa Convention banning the use, stockpiling, production, and transfer of APLs was signed by 122 states in December 1997. The states parties and NGOs then mounted a pressure campaign on holdout states to join the APL ban. The Ottawa Convention entered into force in March 1999, and more than three-quarters of the world's states have ratified it. A new norm has emerged in international relations whereby the possession, trade, and deployment of APLs have become taboo.

The middle powers used their soft power effectively during the negotiations to establish the ICC. Canada created the Like-Minded Group of Countries in 1994 to harness the diplomatic power of the pro-ICC states. The LMG formed a close alliance with the NGO Coalition for an International Criminal Court, which tapped into the energy of the grassroots. Diplomats from the Netherlands and Canada served as chairmen of the Ad Hoc and Preparatory Committees, as well as the Committee of the Whole during the 1998 Rome Conference. They appointed members of the LMG as issue coordinators, thereby ensuring that the pro-ICC bloc would control the negotiations. The LMG adopted the position that the ICC should be free from Security Council oversight, and that the ICC Prosecutor should have the power to initiate investigations of genocide, crimes against humanity, war crimes, and the crime of aggression *proprio motu*. Toward the end of the Rome Conference, LMG members revised the draft statute in private to prevent any attempts at sabotage from ICC skeptics, and then convinced holdout states to vote for the ICC. The Rome Statute was adopted by an overwhelming majority of states, and the powerful ICC that the LMG envisioned became a reality in 2002.

On the issue of regulating the legal trade in SALW, the middle powers adopted a different strategy. Rather than go it alone, employ fast-track diplomacy, or assume control of multilateral negotiations, the middle powers relied on the traditional method of consensus decision-making at the 2001 UN conference and subsequent meetings on SALW. Why would the middle powers return to a traditional method which produces substandard results, considering how effective their previous campaigns were? A possible reason is the absence of a critical mass of support. Unlike the APL and ICC issues, there is no strong majority of states in favor of stricter regulations on the legal SALW trade. Most states depend on the legal trade in order to equip their military forces, and many countries have a SALW industry which generates economic incentives for states to facilitate the legal SALW trade rather than curtail it. Moreover, the global NGO community is divided between anti-SALW and pro-firearms groups. While IANSA lobbies for greater restrictions on SALW, hunting and sport shooting organizations campaign for the protection of citizens' rights to own, purchase, and sell small arms. In short, resistance to the idea of including the legal trade in SALW on the agenda of the UN conference was so strong as to possibly jeopardize any agreements reached on the illicit aspects of the SALW trade. The middle powers realized the importance of achieving an accord on the illicit trade, and therefore participated in the adoption by consensus of the Program of Action, even though the issue of the legal trade was excluded from the PoA.

The middle powers also utilized consensus decision-making on the R2P initiative. Canada took the lead in creating the ICISS, which was co-chaired by the Australian Gareth Evans and featured two Canadian Commissioners. Following the 2001 release of the ICISS report, Canada stood nearly alone among states in calling for international discussion of R2P. In order to generate momentum for R2P, the Canadian government networked with

NGOs, built a coalition of like-minded states, and derived a watered down "R2P lite" (which emphasized that the Security Council should be the exclusive authorizer of interventions) that it promoted as a more palatable alternative to the concept of R2P as proposed by the ICISS (which assigned the Security Council the primary, but not the sole, authority). "R2P lite" proved to be popular among the participants at the 2005 World Summit, though numerous states expressed their skepticism and displeasure. The UN ambassadors from Australia and Canada were instrumental in ensuring that R2P was included in the Outcome Document, despite attempts by R2P skeptics to either scale down the concept further or scuttle it completely. In the case of R2P, the middle powers relied on consensus decision-making with the objective of fostering a widespread acceptance of the responsibility to protect that would hopefully lead to an international norm of R2P in the future.

To summarize, the middle powers have made deliberate choices of diplomatic strategy in order to achieve their objectives. In order to create a rapidly deployable brigade for UN peacekeeping that would remain independent from great power influence and manipulation, the middle powers declined to invite the United States or any of the other P5 states to participate in the formation or deployment of SHIRBRIG, even though the brigade would have benefited from great power assistance in the areas of funding, troop and equipment donations, and logistics. As norm entrepreneurs, the middle powers used fast-track diplomacy, a "take it or leave it" approach, to rapidly generate an effective international norm banning APLs that would not become watered down through exemptions. The middle powers used their soft power resources to maneuver into positions of influence at the Rome Conference, thereby permitting them to control the negotiation process and ensure that no obstructionist states would foil the establishment of an ICC with an independent Prosecutor. On the SALW and R2P initiatives, the middle powers resorted to the consensus decision-making of UN conferences. In the case of SALW, the middle powers realized that reaching an agreement on a PoA to combat the illicit SALW trade was so vitally important that they were unwilling to jeopardize the negotiations by pressing too hard for less popular restrictions on the legal trade. The middle powers were willing to subject the R2P concept to difficult, consensus-based negotiations in order to produce a unanimous agreement that would cement R2P as an international principle. In contrast to the APL ban, the middle powers were content this time with a weaker R2P principle that has broader support among states, and may eventually evolve into a R2P norm as states adapt their behavior to conform to it.

The reactions of the United States

The United States has reacted in different ways to these human security initiatives, based on how the American government has perceived their impact on the US national interest. The creation of SHIRBRIG corresponded with American interest in UN peacekeeping reform. The middle powers did not invite the United States to participate in this initiative, and Washington never expressed its position regarding SHIRBRIG specifically. But the United States publicly supported the idea of a rapidly deployable brigade for UN peacekeeping, as long as it was a standby force rather than a standing army. SHIRBRIG addressed this need, therefore it is safe to assume that Washington approved.

Although the United States was the first country to introduce a moratorium on the export of APLs, it rejected the Ottawa Process because the proposed APL ban would not exempt the American landmines deployed to protect South Korea. The United States preferred to negotiate an APL ban treaty within the channels of the Conference on Disarmament in the belief that the CD would be a more effective forum for convincing major APL producers like China and Russia to join the ban. But to the dismay of the United States the CD would not address the issue of APLs. With the urging of Congress, the United States finally joined the Ottawa Process in August 1997. The Clinton administration proposed several revisions to the draft treaty at the September 1997 Oslo conference, including an exemption for American APLs in Korea. But with the exception of stronger verification measures, the US proposals were rejected. In response, the Clinton administration did not sign the Ottawa Convention banning APLs. Subsequent presidential administrations have also refused to join the Ottawa Convention. But the United States has adopted policies of nonuse, non-export, and nonproduction of APLs, and has no intention of procuring APLs in the future, therefore it may be concluded that Washington has acquiesced to the new international norm banning APLs.

The United States favored the establishment of an ICC, as long as the court would be subjected to Security Council oversight (i.e. it would not have an independent capability to prosecute), and the court would take into account the special responsibilities of the United States for maintaining world peace (i.e. it would exempt from prosecution American military personnel who serve abroad). With these measures in place, the United States would be able to veto any potential prosecutions it disagreed with (as would the rest of the P5), and would be able to protect the constitutional rights of American citizens, under the Fifth and Sixth Amendments, to a jury trial and due process. During the meetings of the Preparatory Committee for the Rome Conference, the US delegation played an active role in defining the crimes that the ICC would have jurisdiction over and in ensuring that due process would be protected in the statute.

But at the Rome Conference, the LMG used its control over the negotiation process to establish an independent ICC Prosecutor who does not require a referral from the Security Council, much to the chagrin of the United States Moreover, the United States objected to the Rome Statute's provision that states not parties may still be subject to potential prosecution for any future crimes that would be added to the statute, even though states parties could opt out of the ICC jurisdiction for new crimes. The American delegation attempted to make the consent of the national government of the accused mandatory for prosecution, but the LMG managed to block this proposition. The United States responded by voting against the ICC. Although President Clinton would later sign the Rome Statute in order to gain some influence over the future development of the court, the Bush administration adopted a hostile position, withdrew American participation, and threatened states parties. But the US policy toward the ICC has mellowed recently, due to the Obama administration's satisfaction with the ICC's professionalism (and disinterest to prosecute US service members, as originally feared). While the United States remains a nonparty to the Rome Statute, it has adopted a more constructive approach of positive engagement with the court.

The US government was highly antagonistic to the notion of instituting stronger regulations on the legal trade in SALW. During the meetings of the Preparatory Committee for the 2001 UN conference on the illicit SALW trade, Canada submitted a paper which stressed the need for any agreement on the illegal trade to also take into account the consequences of the legal trade. The United States was firm in arguing that it would oppose any legally binding treaty that would prevent Americans from engaging in the legal trade of SALW, or would infringe on the constitutional right of American citizens, under the Second Amendment, to own firearms. The American policy during the administrations of both Bill Clinton and George W. Bush was that the United States would only accept a non-legally binding Program of Action that is restricted to dealing with the illicit aspects of the SALW trade.

At the July 2001 UN conference, the US delegation reaffirmed its commitment to eliminating the illicit trade in military-type SALW. But the United States also emphasized that hunting and sport shooting are part of the American culture and that it would protect the gun rights of American citizens. Since the conference featured consensus decision-making, the US delegation was able to ensure that the PoA would be solely a politically binding agreement that would exclude any references to the regulation of private gun ownership, as well as a ban on SALW transfers to nonstate actors and an instrument to enable the tracing of the lines of supply of SALW. With regards to the current negotiations on a legally binding Arms Trade Treaty that would cover SALW, the Obama administration has reversed the policy of the Bush administration and expressed its support, provided that decisions will be made on the basis of consensus to avoid the creation of "loopholes."

In the case of the R2P initiative, the Bush administration rejected the ICISS report as well as any commitments to automatically intervene in the event of mass atrocities. But the administration felt strong pressure from American citizens to halt the genocide in Darfur, and the 2005 Gingrich-Mitchell report endorsed R2P. At the World Summit, the Bush administration adopted the position of the Gingrich-Mitchell report that the R2P of the international community should be secondary to the responsibility of individual states. The US delegation insisted that while R2P may entail a moral responsibility to protect, it should not become a legally binding one. Moreover, the United States wished to maintain its freedom to decide on an ad hoc basis how it would respond to a humanitarian crisis, rather than follow fixed criteria for an intervention. Each of these points, which were of concern not only to the United States but to several other states as well, was ultimately included in the World Summit Outcome Document. The United States joined 173 other states in adopting the Outcome Document, which included the concept of R2P. The Bush administration acquiesced to the World Summit's version of R2P, which was significantly modified from the R2P as proposed by the ICISS, because it addressed the US interest in genocide prevention while incorporating the American points of concern.

Based on these case studies, it appears that the United States is likely to oppose a middle-power-led human security initiative if it perceives the initiative as a threat to the core American national interest, defined as the security of the US territory, institutions, and citizenry. The SHIRBRIG and R2P initiatives corresponded to American interests, and the United States acquiesced to both. The United States was in favor of a rapid response capability for UN peacekeeping, as long as a standby brigade was formed rather than a standing UN army. Moreover, Washington wished to prevent genocide and other mass atrocities from occurring, as long as no international mechanism was established that would force the Americans to intervene automatically in countries of limited interest to the United States The APL ban challenged US security interests abroad but not the core American national interest. Although the Ottawa Convention did not exempt the American landmines in Korea, the United States was already committed to the nonuse, nonproduction, and nontrade of APLs. Thus, the United States rejected the Ottawa Convention, but complied with the new international norm against APLs. In contrast, both the ICC and anti-SALW initiatives were perceived by Washington as posing threats to specific rights of the American citizenry, which are protected under the US Constitution. The United States responded with attempts to defeat these initiatives, unsuccessfully with regards to the creation of the ICC, but successfully in squashing the adoption of restrictions on the legal trade in SALW.

The results of the human security initiatives

It would be overly simplistic to appraise the human security initiatives dichotomously as either a complete success or a total failure. There are definitely shades of gray between the black and white.

The SHIRBRIG initiative provided the UN with a rapid reaction force for peacekeeping missions. The brigade was deployed with considerable success as part of UNMEE, and it participated in a few other missions on the African continent. But while SHIRBRIG addressed a shortcoming in UN peacekeeping, namely the lack of a rapid response capability, it may have already been outdated by the time it was created. The demand for traditional peacekeeping decreased significantly after the Cold War ended, and

SHIRBRIG was not designed for the task of peace enforcement that the UN really needed in the contemporary era. Thus, SHIRBRIG was shut down permanently in 2009 due to a lack of suitable missions, once again leaving the UN without a rapid reaction brigade should a need for traditional peace-keeping ever arise in the future.

The Ottawa Process appears to be the middle powers' most impressive accomplishment. Within a mere fourteen months, a treaty banning APLs was negotiated and signed. The global trade in APLs has become practically negligible, and many states not parties and nonstate actors have even pledged to abide by the Ottawa Convention. The middle powers have clearly played effective roles of norm entrepreneurs in developing and promoting a strong international norm prohibiting APLs. But the three greatest military powers of the world—the United States, Russia, and China—remain nonsignatories of the convention and continue to possess large APL stockpiles, as do several regional powers. APLs are still being manufactured in three countries, and civilians around the world are still being maimed and killed by these indiscriminate weapons.

The creation of the International Criminal Court has also been hailed as a resounding success. The Rome Statute established an ICC that is free from Security Council oversight, with a Prosecutor who can launch investigations independently. The ICC has opened investigations of mass atrocities in six countries, including one *proprio motu*, and trials of accused criminals have begun. But the trials have endured far longer than expected, and the ICC has only concluded one trial and convicted a single perpetrator so far. Moreover, there is less support for the ICC than for the Ottawa Convention, as more than seventy countries have not signed the Rome Statute.

The attempt by the middle powers to derive stricter regulations on the legal trade in SALW has been a flop, at least so far. The hegemonic US was able to ensure that the 2001 New York conference's Program of Action did not include measures to restrict private gun ownership, a prohibition on SALW transfers to nonstate actors, or an instrument to trace the SALW lines of supply. No progress on the legal SALW trade was made at the subsequent biennial meetings and the 2006 Review Conference. But there is some promise that the middle powers' efforts may not be in vain, as diplomatic negotiations have begun on an Arms Trade Treaty that would introduce a legally binding commitment to control the legal flow of SALW. Therefore, it may be premature to classify this human security initiative as an abject failure.

There is a widespread consensus that the R2P initiative was successful in generating a new principle of international relations that emphasizes the responsibility of each state to protect its citizens and refrain from engaging in mass atrocities. The international community agreed overwhelmingly to this redefinition of sovereignty at the 2005 World Summit. The principle of R2P that they endorsed, however, was modified considerably from the original concept that the ICISS formulated and presented in its 2001 report. Ironically the Canadian government, which established the ICISS, ended up watering down the Commission's recommendations into a "R2P lite" that would be more widely acceptable to the international community. Moreover, there is a possibility that the World Summit participants only paid lip service to the notion of R2P. What if the international community ignores its responsibility to protect in the event of future genocide and mass atrocities? Can an unenforced principle be genuinely called a success?

Nevertheless, these five cases of human security initiatives were characterized by significant efforts at leadership by the middle powers. In each of these campaigns, the middle powers sought out like-minded states with which they could form a coalition that would work for the fulfillment of the initiative. With the exception of the SHIRBRIG project, these coalitions included the participation of NGOs, which could harness the energies of a mobilized civil society in support of the human security goal. Whether or not they have been successful in achieving their objectives, it is clear that the middle powers have exercised considerable influence on issues of global security.

The future of human security

Human security certainly has its share of critics. One of the most cited critiques is a 2001 article by Roland Paris in *International Security*. Paris acknowledged that human security is useful as an all-inclusive campaign slogan under which middle powers, international organizations, and NGOs can unite their efforts to achieve particular objectives on behalf of humanity. But human security has limited utility for scholars and foreign policy-makers, due to the vaguely defined and all-encompassing nature of this concept. As Paris stated, "if human security means almost anything, then it effectively means nothing" (Paris 2001, 93).

Taylor Owen (2008) argued that the term "human security" is often used interchangeably with the terms "human development" and "human rights" in the UN discourse, thereby creating confusion as to whether or not human security is actually making unique theoretical and practical contributions. Moreover, human security suffers from conceptual overstretch, where it has lost most of its analytical value by being overly inclusive (MacFarlane and Khong 2006; Owen 2008). The consequences of securitizing more issues include the creation of false priorities on the security agenda, causal confusion with regards to variables, and the possibility that governments will rely on military remedies that would only exacerbate human insecurity.

Mandy Turner, Neil Cooper, and Michael Pugh made the claim that human security cannot become an emancipatory approach "because it has been institutionalised and co-opted to work in the interests of global capitalism, militarism and neoliberal governance" (Turner, Cooper, and Pugh 2011, 83). Amitav Acharya (2001) explained that Asian governments have been warm toward the notion of human security defined as freedom from want, because it is compatible with the popular Asian conceptualization of comprehensive security, which recognizes both military and nonmilitary threats to the state. But governments in Asia have been skeptical about the freedom from fear perspective, viewing this as a veiled attempt by the West to impose liberal values and institutions in the region. Moreover, human security is perceived as an ideological tool that is wielded by "like-minded" states and is possibly divisive for attempts to achieve cooperative security in Asia.

T. S. Hataley and Kim Richard Nossal (2004) indicated that while the academic community has been devoting more attention to the concept of human security, national governments have been reluctant to adopt policies that are geared toward ensuring human security. The Canadian government is not immune from this charge. For example, Canada was very slow in assembling a nominal peacekeeping contingent for East Timor in 1999, due to a minimal interest in East Timor rather than a genuine interest in the human security of the East Timorese. David Bosold and Wilfried Von Bredow (2006) discovered that despite its human security rhetoric, Canadian foreign policy has not become more idealistic over time, and its issue agenda has not broadened significantly. Rather, the types of issues which Canada has addressed, and the methods it has used, have been consistent with Canadian foreign policy in the past. In fact, "human security and soft power only complement other areas and forms of foreign policy. Hence, human security is not a replacement of security issues such as nonproliferation, but rather a supplement to it" (Bosold and Von Bredow 2006, 843). In their study of Canadian and Japanese foreign policies regarding the Balkans, Asteris Huliaras and Nikolaos Tzifakis concluded that "from a realist perspective, the Japanese and—to a lesser extent—the Canadian human security approaches were intended merely to lend new vigour and effectiveness to existing policies and thus they were equivalent to little more than refurbishments of previous instruments and policies" (Huliaras and Tzifakis 2007, 574).

To summarize, it appears that human security may not be the new theoretical paradigm to displace realism that critics of power politics have been craving for so long. Problems of vague definitions, all inclusiveness, and conceptual overstretch plague the human security approach. In addition, skeptics are wary of the human security agenda, believing that it is merely a rhetorical masquerade for the ideological, economic, and geopolitical interests of the like-minded Western middle powers. Studies have indicated that despite the proclamation of the human security agenda, the middle powers have not transformed their foreign policies in significant ways. Rather, middlepowermanship has continued to use the same methods and emphasize similar issues as before. Realpolitik remains the dominant driving force behind middle power foreign policies, rather than considerations of the human interest beyond national borders. Symbolic of the possibly fleeting commitment of the middle powers to human security is the fact that the website of the Human Security Network is no longer accessible on the internet.

But despite the serious criticisms, the concept of human security has made positive contributions to international relations. Human security serves as a conceptual umbrella under which like-minded states, international organizations, and NGOs come together to discuss issues that affect humanity. It is also a rallying call which motivates actors to not accept human insecurity as a natural condition of our world. Although human security has deficiencies as a theoretical paradigm, it does provide a solid framework for analyzing the extent to which numerous phenomena, of both natural and human origin, threaten the security of human populations. I believe that the greatest contribution of the human security approach is informative, as it shines a light on the numerous ways in which human populations are endangered, which traditional security perspectives fail to do.

To address the concerns of Steven Walt (1991) and other traditionalists who worry that an expansion of the security agenda would destroy the intellectual coherence of security studies, I suggest that human security be viewed as an interdisciplinary endeavor rather than as a subfield of security studies. Each of the UNDP's (1994) subcategories of human security—economic well-being, food security, health, environmental conservation, personal security, community security, and political security—has its own scholars and practitioners who have the necessary expertise to address particular issues and derive solutions. Political scientists need only concern themselves with issues that lie within the areas of community and political security, as scholars in other disciplines will be better equipped to deal with other aspects of human security. International relations specialists can contribute what they do best: analyze the effects that international and transnational cooperation, conflict, institutions, trade, communications, migration, and so forth have on human security. Any holistic assessments of human security, whether at the global, regional, or country level, should involve interdisciplinary cooperation between scholars.

In conclusion, this book demonstrated that middle powers are capable of playing leadership roles in the domain of global security. While the findings may surprise many international relations scholars who have been raised on a diet of realism and great power politics, they should provide support and encouragement for foreign policy officials in middle power countries who have designed and implemented initiatives which make the lives of civilians around the world more secure. It is possible for them to build a safer and better world after all.

REFERENCES

Acharya, Amitav. 2001. "Human Security: East versus West." *International Journal* 56(3): 442–60.

Advisory Council on International Affairs. 2010. *The Netherlands and the Responsibility to Protect: The Responsibility to Protect People From Mass Atrocities*. No. 70, June. The Hague: Advisory Council on International Affairs.

Albright, Madeleine K. 1993. "Building a Consensus on International Peace-Keeping." *US Department of State Dispatch* 4(46): 789–92.

Aldinger, Charles. 2004. "U.S. Removes Peacekeepers Over War Crimes Court." *Reuters*, 1 July. <http://news.yahoo.com/news?tmpl=story&u=/nm/20040701/us_nm/un_usa_troops_dc_4>(Accessed 3 July 2004).

American Medical Association. 1997. "Landmine-Related Injuries, 1993–96." *JAMA, The Journal of the American Medical Association* 278(8): 621.

American Non-Governmental Organizations Coalition for the International Criminal Court. 2002a. "Bilateral Immunity Agreements." <http://www.amicc.org/usinfo/administration_policy_BIAs.html#countries> (Accessed on 21 February 2007).

—2002b. "Congressional Update." <http://www.amicc.org/usinfo/congressional.html> (Accessed on 21 February 2007).

—2004. *Information about the US and the ICC*. New York: American Non-Governmental Organizations Coalition for the International Criminal Court.

American Society of International Law. 2006. "U.S. Officials Endorse 'Responsibility to Protect' through Security Council Action." *The American Journal of International Law* 100(2): 463–4.

—2007. "President and Congress End Limits on Military Training for Parties to ICC Treaty." *The American Journal of International Law* 101(1): 213–15.

—2009. "United States Eases Opposition to International Criminal Court, Opposes Efforts to Thwart ICC Proceedings Involving Darfur." *The American Journal of International Law* 103(1): 152–4.

—2010. "U.S. Delegation Active in ICC Negotiations to Define Crime of Aggression." *The American Journal of International Law* 104(3): 511–14.

Anbarasan, Ethirajah. 1998. "A Decisive Victory." *UNESCO Courier* 51(October): 29–31.

Anderson, John B. 2002. "Unsigning the ICC." *The Nation*, 29 April.

Arms Project of Human Rights Watch and Physicians for Human Rights. 1993. *Landmines: A Deadly Legacy*. New York: Human Rights Watch.

Arsanjani, Mahnoush H. 1999. "The Rome Statute of the International Criminal Court." *The American Journal of International Law* 93(1): 22–43.

Asada, Masahiko. 1995. "Peacemaking, Peacekeeping, and Peace Enforcement: Conceptual and Legal Underpinnings of the U.N. Role." In *U.N. Peacekeeping: Japanese and American Perspectives*, eds. Selig S. Harrison and Masashi Nishihara. Washington, DC: Carnegie Endowment for International Peace, pp. 31–70.

Axworthy, Lloyd. 1997. "Canada and Human Security: The Need for Leadership." *International Journal* 52(2): 183–96.

—2003. *Navigating a New World: Canada's Global Future*. Toronto: Alfred A. Knopf Canada.

Badescu, Cristina G., and Thomas G. Weiss. 2010. "Misrepresenting R2P and Advancing Norms: An Alternative Spiral?" *International Studies Perspectives* 11(4): 354–74.

Barash, David P., ed. 2010. *Approaches to Peace: A Reader in Peace Studies*. Second Edition. New York: Oxford University Press.

Bassiouni, Cherif. 2009. "Advancing the Responsibility to Protect Through International Criminal Justice." In *Responsibility to Protect: The Global Moral Compact for the 21st Century*, eds. Richard H. Cooper and Juliette Voïnov Kohler. New York: Palgrave Macmillan, pp. 31–42.

Baxter, Richard R. 1977. "Conventional Weapons under Legal Prohibitions." *International Security* 1(3): 42–61.

Baylis, John, Ken Booth, John Garnett, and Phil Williams, eds. 1987. *Contemporary Strategy I: Theories and Concepts*. Second Edition. New York: Holmes and Meier Publishers.

Beard, Charles A. 1934. *The Idea of National Interest: An Analytical Study in American Foreign Policy*. New York: The Macmillan Company.

Behringer, Ronald M. 2005. "Middle Power Leadership on the Human Security Agenda." *Cooperation and Conflict* 40(3): 305–42.

Bellamy, Alex J. 2009a. "Realizing the Responsibility to Protect." *International Studies Perspectives* 10(2): 111–28.

—2009b. *Responsibility to Protect: The Global Effort to End Mass Atrocities*. Cambridge: Polity Press.

—2010a. "The Responsibility to Protect and Australian Foreign Policy." *Australian Journal of International Affairs* 64(4): 432–48.

—2010b. "The Responsibility to Protect—Five Years On." *Ethics & International Affairs* 24(2): 143–69.

Benjamin, Daniel. 2001. "Relaxing the Strong Arm: The U.S. May Not Want a Global War-Crimes Court, But Why Spoil It For Others?" *Time International*, 10 September.

Berg, Karen. 1997. "A Permanent International Criminal Court." *UN Chronicle* 34(4): 30–6.

Berridge, G. R. 2005. *Diplomacy: Theory and Practice*. Third Edition. Basingstoke, UK: Palgrave Macmillan.

"Big Damage; Small Arms; The Illicit Trade in Light Weapons." 2001. *The Economist*, 14 July.

"The Birth of a New World Court." 1998. *Maclean's*, 27 July.

Booth, Ken, ed. 2005. *Critical Security Studies and World Politics*. Boulder, CO: Lynne Rienner.

Bosco, David L. 1998. "Sovereign Myopia." *The American Prospect* 41 (November–December): 24–7.
Bosold, David, and Wilfried Von Bredow. 2006. "Human Security: A Radical or Rhetorical Shift in Canada's Foreign Policy?" *International Journal* 61(4): 829–44.
"Both Sides Lose; The International Criminal Court." 2002. *The Economist*, 20 July.
Boutros-Ghali, Boutros. 1994. "The Land Mine Crisis: A Humanitarian Disaster." *Foreign Affairs* 73(5): 8–13.
Boutwell, Jeffrey, and Michael T. Klare. 1999. "Introduction." In *Light Weapons and Civil Conflict: Controlling the Tools of Violence*, eds. Jeffrey Boutwell and Michael T. Klare. Lanham, MD: Rowman & Littlefield, pp. 1–5.
Bring, Ove. 1987. "Regulating Conventional Weapons in the Future. Humanitarian Law or Arms Control?" *Journal of Peace Research* 24(3): 275–86.
Brown, Bartram S. 2000. "The Statute of the ICC: Past, Present, and Future." In *The United States and the International Criminal Court: National Security and International Law*, eds. Sarah B. Sewall and Carl Kaysen. Lanham, MD: Rowman & Littlefield, pp. 61–84.
Brown, Michael E. 2003. "Security Problems and Security Policy in a Grave New World." In *Grave New World: Security Challenges in the 21st Century*, ed. Michael E. Brown. Washington, DC: Georgetown University Press, pp. 305–27.
Burchill, Scott. 2005. *The National Interest in International Relations Theory*. Basingstoke, UK: Palgrave Macmillan.
Buzan, Barry. 1983. *People, States and Fear: The National Security Problem in International Relations*. First Edition. Brighton, UK: Wheatsheaf.
—1991. *People, States and Fear: An Agenda for International Security Studies in the Post-Cold War Era*. Second Edition. Boulder, CO: Lynne Rienner.
—1997. "Rethinking Security after the Cold War." *Cooperation and Conflict* 32(1): 5–28.
Buzan, Barry, and Lene Hansen. 2009. *The Evolution of International Security Studies*. Cambridge: Cambridge University Press.
Buzan, Barry, Ole Wæver, and Jaap de Wilde. 1998. *Security: A New Framework for Analysis*. Boulder, CO: Lynne Rienner.
Cameron, Maxwell A. 2002. "Global Civil Society and the Ottawa Process: Lessons From the Movement to Ban Anti-Personnel Mines." In *Enhancing Global Governance: Towards A New Diplomacy?*, eds. Andrew F. Cooper, John English, and Ramesh Thakur. New York: United Nations University Press, pp. 69–89.
Canada, Department of Foreign Affairs and International Trade. 1995. *Towards a Rapid Reaction Capability for the United Nations*. Ottawa: Department of Foreign Affairs and International Trade, September.
—2000. *Freedom from Fear: Canada's Foreign Policy for Human Security*. Ottawa: Department of Foreign Affairs and International Trade.
—2002. *Freedom from Fear: Canada's Foreign Policy for Human Security*. Second Edition. Ottawa: Department of Foreign Affairs and International Trade.
Canada. Permanent Mission of Canada to the United Nations. 2003. *Statement by Robert McDougall, Head of the Canadian Delegation, at the First Biennial Meeting of States on the Implementation of the Programme of Action of the 2001 United Nations Conference on the Illicit Trade in Small Arms and Light Weapons*. New York, 7 July.

—2005. *Statement by Mr. Tim Martin, Head of the Canadian Delegation, to the Second Biennial Meeting of States on the Implementation of the Programme of Action of the 2001 United Nations Conference on the Illicit Trade in Small Arms and Light Weapons*. New York, 11 July.

Canada. 2010. *Preventing and Combating the Illicit Trade in SALW across Borders*. Statement by Canada at the Biennial Meeting of States on the UN Programme of Action to Prevent, Combat and Eradicate the Illicit Trade in Small Arms and Light Weapons in All Its Aspects, New York, 14–18 June.

Capellaro, Catherine, and Anne-Marie Cusac. 1997. "Meet The People Who Make Land Mines." *The Progressive* 61(11): 18–22.

Carr, Edward H. [1939] 1964. *The Twenty Years' Crisis 1919–39: An Introduction to the Study of International Relations*. New York: Harper & Row, Publishers.

Carter, Gregg L. 1997. *The Gun Control Movement*. New York: Twayne.

Carter, Tom. 2002. "World Tribunal a Done Deal; the International Criminal Court is Four Signatories Short of Ratification, Which Could Happen This April." *Insight on the News* 18(15): 29.

Casey, Lee A., and David B. Rivkin, Jr. 2003. "The Rocky Shoals of International Law." In *The National Interest on International Law & Order*, ed. R. James Woolsey. New Brunswick, NJ: Transaction, pp. 3–15.

Cassel, Doug. 2001. "Bad Neighbors." *The American Prospect* 12(15): 14.

Clegg, Liz. 1999. "NGOs Take Aim." *Bulletin of the Atomic Scientists* 55(1): 49.

Cocklin, Chris. 2002. "Water and 'Cultural Security'." In *Human Security and the Environment: International Comparisons*, eds. Edward A. Page and Michael Redclift. Cheltenham, UK: Edward Elgar, pp. 154–76.

Cohen, Roberta, and Francis M. Deng. 1998. *Masses in Flight: The Global Crisis of Internal Displacement*. Washington, DC: The Brookings Institution.

Colwill, Jeremy. 1995. "From Nuremberg to Bosnia and Beyond: War Crimes Trials in the Modern Era." *Social Justice* 22(3): 111–28.

"The Constitution of the United States of America." [1787] 2002. In *The Founders' Almanac: A Practical Guide to the Notable Events, Greatest Leaders & Most Eloquent Words of the American Founding*, Reference Edition, ed. Matthew Spalding. Washington, DC: Heritage Foundation, pp. 245–79.

Cooper, Andrew F. 1992. "Multilateral Leadership: The Changing Dynamics of Canadian Foreign Policy." In *Making A Difference? Canada's Foreign Policy in a Changing World Order*, eds. John English and Norman Hillmer. Toronto: Lester, pp. 200–21.

—1997a. "Niche Diplomacy: A Conceptual Overview." In *Niche Diplomacy: Middle Powers after the Cold War*, ed. Andrew F. Cooper. New York: St. Martin's, pp. 1–24.

Cooper, Andrew F., ed. 1997b. *Niche Diplomacy: Middle Powers after the Cold War*. New York: St. Martin's.

Cooper, Andrew F., Richard A. Higgott, and Kim Richard Nossal. 1993. *Relocating Middle Powers: Australia and Canada in a Changing World Order*. Vancouver: UBC Press.

Cordner, Stephen, and Helen McKelvie. 1998. "Forensic Medicine: International Criminal Tribunals and an International Criminal Court." *The Lancet*, 27 June.

Cox, David. 1993. *Exploring An Agenda for Peace: Issues Arising from the Report of the Secretary-General*. Ottawa: Canadian Centre for Global Security.

Cox, Robert W. 1981. "Social Forces, States, and World Orders: Beyond International Relations Theory." *Millennium: Journal of International Studies* 10(2): 126–55.

—1989. "Middlepowermanship, Japan, and Future World Order." *International Journal* 44(4): 823–62.

Cox, Robert W., and Harold K. Jacobson. 1973. *The Anatomy of Influence: Decision Making in International Organization*. New Haven, CT: Yale University Press.

Crook, John R. 2010. "United States Sends Observers to ICC Assembly of States Parties." *The American Journal of International Law* 104(1): 126–7.

"Curbing Horror; Landmines." 2001. *The Economist*, 29 September.

Daniel, Donald C. F., and Bradd C. Hayes with Chantal de Jonge Oudraat. 1999. *Coercive Inducement and the Containment of International Crises*. Washington, DC: United States Institute of Peace.

Davies, Paul, with photographs by Nic Dunlop. 1994. *War of the Mines: Cambodia, Landmines and the Impoverishment of a Nation*. London: Pluto.

"The Declaration of Independence." [1776] 2002. In *The Founders' Almanac: A Practical Guide to the Notable Events, Greatest Leaders & Most Eloquent Words of the American Founding*, Reference Edition, ed. Matthew Spalding. Washington, DC: Heritage Foundation, pp. 221–6.

Deng, Francis M., Sadikiel Kimaro, Terrence Lyons, Donald Rothchild, and I. William Zartman. 1996. *Sovereignty as Responsibility: Conflict Management in Africa*. Washington, DC: The Brookings Institution.

Denmark, Chief of Defence. 1995a. *United Nations Standby Arrangements for Peacekeeping: A Multinational UN Standby Forces High Readiness Brigade*. 25 January.

—1995b. *Report by the Working Group on a Multinational UN Standby Forces High Readiness Brigade*. 15 August.

Dhanapala, Jayantha. 2002. "Multilateral Cooperation on Small Arms and Light Weapons: From Crisis to Collective Response." *The Brown Journal of World Affairs* 9(1): 163–71.

Diehl, Paul F. 1993. *International Peacekeeping*. London: Johns Hopkins University Press.

Dueck, Colin. 2005. "Realism, Culture and Grand Strategy: Explaining America's Peculiar Path to World Power." *Security Studies* 14(2): 195–231.

Eavis, Paul, and William Benson. 1999. "The European Union and the Light Weapons Trade." In *Light Weapons and Civil Conflict: Controlling the Tools of Violence*, eds. Jeffrey Boutwell and Michael T. Klare. Lanham, MD: Rowman & Littlefield, pp. 89–100.

Elgström, Ole. 1992. *Foreign Aid Negotiations: The Swedish-Tanzanian Aid Dialogue*. Aldershot, UK: Avebury.

Elwell, Christine. 1998. "New Trade and Environmental Compliance Measures to Enhance Conventional Arms Agreements: From Landmines to UN Peace-keeping." In *Treaty Compliance: Some Concerns and Remedies*, ed. Canadian Council on International Law. London: Kluwer Law International, pp. 35–86.

European Union. Italian Presidency of the Council of the European Union. 2003. *Biennial Meeting of States to Consider the Implementation of the Programme of Action on Small Arms, Statement by H. E. Ambassador Carlo Trezza, Permanent*

Representative of Italy to the Conference on Disarmament in Geneva on behalf of the European Union. New York, 7 July.

Evans, Gareth. 2009. "The Responsibility to Protect: From an Idea to an International Norm." In *Responsibility to Protect: The Global Moral Compact for the 21st Century*, eds. Richard H. Cooper and Juliette Voïnov Kohler. New York: Palgrave Macmillan, pp. 15–29.

Everett, Robinson O. 2000. "American Servicemembers and the ICC." In *The United States and the International Criminal Court: National Security and International Law*, eds. Sarah B. Sewall and Carl Kaysen. Lanham, MD: Rowman & Littlefield, pp. 137–51.

Eviatar, Daphne. 2001. "Mugging the ICC." *The Nation*, 5 November.

Faulkner, Frank. 1997. "The Most Pernicious Weapon: Landmines." *Contemporary Review* 270(1574): 136–42.

—1998. "Some Progress on Landmines." *Contemporary Review* 273(1590): 1–5.

Feinstein, Lee, and Erica De Bruin. 2009. "Beyond Words: U.S. Policy and the Responsibility to Protect." In *Responsibility to Protect: The Global Moral Compact for the 21st Century*, eds. Richard H. Cooper and Juliette Voïnov Kohler. New York: Palgrave Macmillan, pp. 179–98.

Ferencz, Benjamin B. 1980. *Half a Century of Hope*. Vol. 1 of *An International Criminal Court, A Step Toward World Peace – A Documentary History and Analysis*. New York: Oceana Publications.

Fields, Felicity. 2001. "A World Without Landmines." *The Humanist* 61(5): 13.

Finnemore, Martha, and Kathryn Sikkink. 1998. "International Norm Dynamics and Political Change." *International Organization* 52(4): 887–917.

Florini, Ann. 1996. "The Evolution of International Norms." *International Studies Quarterly* 40(3): 363–89.

Fortier, Patricia. 2001. "The Evolution of Peacekeeping." In *Human Security and the New Diplomacy: Protecting People, Promoting Peace*, eds. Rob McRae and Don Hubert. Montreal & Kingston: McGill-Queen's University Press, pp. 41–54.

Frankel, Joseph. 1970. *National Interest*. New York: Praeger.

Frieden, Jeff. 1988. "Sectoral Conflict and Foreign Economic Policy, 1914–40." *International Organization* 42(1): 59–90.

Frye, Alton. 1999. *Toward an International Criminal Court?: Three Options Presented as Presidential Speeches*. New York: Council on Foreign Relations.

Garcia, Denise. 2002. "The Diffusion of Norms in International Security: The Cases of Small Arms and Landmines." Paper presented at the American Political Science Association 98th Annual Convention, Boston, MA, 29 August – 1 September, 2002.

Garner, David. 1997. "Hopes for Mines Ban." *Geographical Magazine* 69(11): 66–71.

Geneva International Centre for Humanitarian Demining. 2005. *From Ottawa to Nairobi and Beyond: Key Documents in the Global Effort to End the Suffering Caused by Anti-personnel Mines*. Geneva: Geneva International Centre for Humanitarian Demining, June.

Gilpin, Robert. 1981. *War and Change in World Politics*. New York: Cambridge University Press.

Gingrich, Newt, and George Mitchell. 2005. *American Interests and UN Reform: A Report of the Congressional Task Force on the United Nations*. Washington, DC: U.S. Institute of Peace.

GlobalSecurity.org. 2006. "The Lord's Resistance Army (LRA)." <http://www.globalsecurity.org/military/world/para/lra.htm> (Accessed 17 February 2007).

Goertz, Gary. 2003. *International Norms and Decision Making: A Punctuated Equilibrium Model*. Lanham, MD: Rowan & Littlefield.

Goldstone, Richard J. 2000. *For Humanity: Reflections of a War Crimes Investigator*. New Haven, CT: Yale University Press.

Goulding, Marrack. 1993. "The Evolution of United Nations Peacekeeping." *International Affairs* 69(3): 451–64.

Gwozdecky, Mark, and Jill Sinclair. 2001. "Case Study: Landmines and Human Security." In *Human Security and the New Diplomacy: Protecting People, Promoting Peace*, eds. Rob McRae and Don Hubert. Montreal & Kingston: McGill-Queen's University Press, pp. 28–40.

Hall, John A., and Charles Lindholm. 1999. *Is America Breaking Apart?* Princeton, NJ: Princeton University Press.

Hampson, Fen Osler, with Jean Daudelin, John B. Hay, Holly Reid, and Todd Martin. 2002. *Madness in the Multitude: Human Security and World Disorder*. Don Mills, ON: Oxford University Press.

Handel, Michael. 1981. *Weak States in the International System*. London: Frank Cass.

Hataley, T. S., and Kim Richard Nossal. 2004. "The Limits of the Human Security Agenda: The Case of Canada's Response to the Timor Crisis." *Global Change, Peace & Security* 16(1): 5–17.

Hay, Robin J. 1999. "Present at the Creation? Human Security and Canadian Foreign Policy in the Twenty-first Century." In *Canada Among Nations 1999: A Big League Player?*, eds. Fen Osler Hampson, Martin Rudner, and Michael Hart. Don Mills, ON: Oxford University Press, pp. 215–32.

Hayes, Geoffrey. 1997. "Canada as a Middle Power: The Case of Peacekeeping." In *Niche Diplomacy: Middle Powers after the Cold War*, ed. Andrew F. Cooper. New York: St. Martin's, pp. 73–89.

Heje, Claus. 1998. "United Nations Peacekeeping – An Introduction." In *A Future for Peacekeeping?*, ed. Edward Moxon-Browne. London: Macmillan, pp. 1–25.

Helleiner, Gerald K., ed. 1990. *The Other Side of International Development Policy: The Non-Aid Economic Relations with Developing Countries of Canada, Denmark, the Netherlands, Norway, and Sweden*. Toronto: University of Toronto Press.

Henrikson, Alan K. 1997. "Middle Powers as Managers: International Mediation Within, Across, and Outside Institutions." In *Niche Diplomacy: Middle Powers after the Cold War*, ed. Andrew F. Cooper. New York: St. Martin's, pp. 46–72.

Higgott, Richard. 1997. "Issues, Institutions and Middle-Power Diplomacy: Action and Agendas in the Post-Cold War Era." In *Niche Diplomacy: Middle Powers after the Cold War*, ed. Andrew F. Cooper. New York: St. Martin's, pp. 25–45.

Hillen, John. 1998. *Blue Helmets: The Strategy of UN Military Operations*. Washington, DC: Brassey's.

Holbraad, Carsten. 1984. *Middle Powers in International Politics*. New York: St. Martin's.

Holmes, John W. 1965. "Is There a Future for Middlepowermanship?" In *Canada's Role as a Middle Power; Papers Given at the Third Annual Banff Conference on World Development, August 1965*, ed. J. King Gordon. Toronto: Canadian Institute of International Affairs, pp. 13–28.

Holsti, Ole R. 1995. "Theories of International Relations and Foreign Policy: Realism and Its Challengers." In *Controversies in International Relations Theory: Realism and the Neoliberal Challenge*, ed. Charles W. Kegley, Jr. New York: St. Martin's, pp. 35–65.

Huldt, Bo. 1995. "Working Multilaterally: The Old Peacekeepers' Viewpoint." In *Beyond Traditional Peacekeeping*, eds. Donald C. F. Daniel and Bradd C. Hayes with Chantal de Jonge Oudraat. New York: St. Martin's, pp. 101–19.

Huliaras, Asteris, and Nikolaos Tzifakis. 2007. "Contextual Approaches to Human Security: Canada and Japan in the Balkans." *International Journal* 62(3): 559–75.

Human Rights Watch. 2002. "U.S.: 'Hague Invasion Act' Becomes Law. White House "Stops at Nothing" in Campaign Against War Crimes Court." 3 August. <http://www.hrw.org/press/2002/08/aspa080302.htm> (Accessed 5 February 2007).

—2003. *Landmine Monitor Fact Sheet: National Implementation Measures (Article 9)*. For the Seventh Meeting of the Intersessional Standing Committee on the General Status and Operation of the 1997 Mine Ban Treaty, Geneva, Switzerland, 7 February. <http://www.icbl.org/content/download/20089/387671/file/legislation_feb_2003.pdf> (Accessed 16 September 2011).

—2009. "US: Obama Should Join Mine Ban Treaty." 25 November. <http://www.hrw.org/en/news/2009/11/25/us-obama-rejection-mine-ban-treaty-reprehensible> (Accessed 23 October 2010).

Human Rights Watch Arms Project and Human Rights Watch/Africa. 1994. *Landmines in Mozambique*. New York: Human Rights Watch.

Humanitarian Coalition on Small Arms. 2001. "Annex A: The U.N. Conference—A Human Rights and Humanitarian Perspective." In *Small Arms and the Humanitarian Community: Developing A Strategy for Action*. Nairobi, Kenya—November 18–20, 2001, Report of the Proceedings. <http://www.iansa.org/documents/2002/Nairobi_Conference_report.htm> (Accessed 26 April 2003).

Hurrell, Andrew. 2000. "Some Reflections on the Role of Intermediate Powers in International Institutions." In *Paths to Power: Foreign Policy Strategies of Intermediate States*, Andrew Hurrell, Andrew F. Cooper, Guadelupe González González, Ricardo Ubiraci Sennes, and Srini Sitaraman. Washington, DC: Latin American Program, Woodrow Wilson Center for Scholars, pp. 1–11.

Hurrell, Andrew, Andrew F. Cooper, Guadelupe González González, Ricardo Ubiraci Sennes, and Srini Sitaraman. 2000. *Paths to Power: Foreign Policy Strategies of Intermediate States*. Washington, DC: Latin American Program, Woodrow Wilson Center for Scholars.

Integrated Regional Information Networks. 2005. "Central African Republic-Chad: Another 4,000 Central Africans Flee to Chad, Recount Village Raids." New York: UN Office for the Coordination of Humanitarian Affairs, 17 August. <http://www.globalsecurity.org/military/library/news/2005/08/mil-050817-irin04.htm> (Accessed 19 February 2007).

—2006. "Cameroon-Central African Republic: Thousands Seek Refuge from Attacks in CAR." New York: UN Office for the Coordination of Humanitarian Affairs, 30 November. <http://www.globalsecurity.org/military/library/news/2006/11/mil-061130-irin03.htm> (Accessed 19 February 2007).

International Action Network on Small Arms. 2003. *Presentations by the International Action Network on Small Arms (IANSA) to the 2003 Biennial Meeting of States on the Implementation of the Programme of Action to Prevent, Combat and Eradicate the Illicit Trade in Small Arms and Light Weapons in All Its Aspects*. New York, 9 July.

—2006. *Gun Violence: A Global Epidemic*. London: International Action Network on Small Arms.

—2011. *Differences between the 'United Nations Programme of Action' and the 'United Nations Arms Trade Treaty'*. London: International Action Network on Small Arms.

International Campaign to Ban Landmines. 2002. "Landmine Monitor Report 2002: Toward a Mine-Free World. Key Developments Since March 2001." <http://www.icbl.org/lm/2002/developments.html> (Accessed 16 September 2011).

—2006. "Landmine Monitor Report 2006. Major Findings." <http://www.icbl.org/lm/2006/findings.html> (Accessed 16 September 2011).

—2009a. "Landmine Monitor Report 2009: Toward a Mine-Free World. Ban Policy." <http://www.the-monitor.org/index.php/publications/display?url=lm/2009/es/ban.html> (Accessed 16 September 2011).

—2009b. "Landmine Monitor Report 2009: Toward a Mine-Free World. Major Findings." <http://www.the-monitor.org/index.php/publications/display?url=lm/2009/es/major_findings.html> (Accessed 16 September 2011).

—2010a. *Cartagena Summit on a Mine-Free World: 29 November–4 December 2009*. ICBL Report on Activities. Geneva: International Campaign to Ban Landmines, May.

—2010b. "Landmine Monitor 2010: Major Findings." <http://www.the-monitor.org/index.php/publications/display?url=lm/2010/es/Major_Findings.html> (Accessed 16 September 2011).

—2011a. "Mine Ban Treaty-States not Parties." <http://www.icbl.org/index.php/icbl/Universal/MBT/States-Not-Party> (Accessed 16 September 2011).

—2011b. "Mine Ban Treaty-States Parties." <http://www.icbl.org/index.php/icbl/Universal/MBT/States-Parties> (Accessed 16 September 2011).

International Coalition for the Responsibility to Protect. 2009. "Statement by Ambassador Susan Rice on Respect for Humanitarian Law in the Security Council." 29 January. <http://www.responsibilitytoprotect.org/index.php/crises/37-the-crisis-in-darfur/2109-us-ambassador-to-the-un-susan-rice-voices-us-support-for-rtop> (Accessed 8 September 2011).

International Commission of Inquiry on Darfur. 2005. *Report of the International Commission of Inquiry on Darfur to the United Nations Secretary-General*. Geneva: International Commission of Inquiry on Darfur, 25 January.

International Commission on Intervention and State Sovereignty. 2001a. *The Responsibility to Protect: Report of the International Commission on Intervention and State Sovereignty*. Ottawa: International Development Research Centre, December.

—2001b. *The Responsibility to Protect: Research, Bibliography, Background. Supplementary Volume to the Report of the International Commission on Intervention and State Sovereignty*. Ottawa: International Development Research Centre, December.

International Criminal Court. 2004a. "President of Uganda Refers Situation Concerning the Lord's Resistance Army (LRA) to the ICC." 29 January. <http://www.icc-cpi.int/pressrelease_details&id=16&l=en.html> (Accessed 3 February 2007).

—2004b. "Prosecutor of the International Criminal Court Opens an Investigation into Northern Uganda." 29 July. <http://www.icc-cpi.int/pressrelease_details&id=33&l=en.html> (Accessed 3 February 2007).

—2004c. "Prosecutor Receives Referral of the Situation in the Democratic Republic of Congo." 19 April. <http://www.icc-cpi.int/pressrelease_details&id=19&l=en.html> (Accessed 3 February 2007).

—2004d. "The Office of the Prosecutor of the International Criminal Court Opens its First Investigation." 23 June. <http://www.icc-cpi.int/pressrelease_details&id=26&l=en.html> (Accessed 3 February 2007).

—2005a. *Facts and Procedure Regarding the Situation in Uganda.* Basic Information, No: ICC20051410.056.1-E. The Hague: International Criminal Court, 14 October.

—2005b. "Prosecutor Receives Referral Concerning Central African Republic." 7 January. <http://www.icc-cpi.int/pressrelease_details&id=87&l=en.html> (Accessed 3 February 2007).

—2005c. *Statement by the Chief Prosecutor on the Uganda Arrest Warrants.* Statement by Luis Moreno-Ocampo, Prosecutor of the International Criminal Court, 14 October 2005. The Hague: International Criminal Court, Office of the Prosecutor.

—2005d. *The Investigation in Northern Uganda.* International Criminal Court, Office of the Prosecutor Press Conference, 14 October 2005. The Hague: International Criminal Court.

—2005e. "The Prosecutor of the ICC Opens Investigation in Darfur." 6 June. <http://www.icc-cpi.int/pressrelease_details&id=107&l=en.html> (Accessed 3 February 2007).

—2006a. *Newsletter* (10). The Hague: International Criminal Court, November.

—2006b. *Situation Democratic Republic of Congo.* Pre-Trial Chamber I, Case Number ICC-01/04-01/06, First Appearance Hearing in Open Session, 20 March 2006. The Hague: International Criminal Court.

—2007a. "Pre-Trial Chamber I Commits Thomas Lubanga Dyilo For Trial." 29 January. <http://www.icc-cpi.int/press/pressreleases/220.html> (Accessed 13 July 2007).

—2007b. "Prosecutor Opens Investigation in the Central African Republic." 22 May. <http://www.icc-cpi.int/pressrelease_details&id=248&l=en.html> (Accessed 12 July 2007).

—2007c. *Request to the Republic of the Sudan for the Arrest and Surrender of Ali Kushayb.* Pre-Trial Chamber I, Case Number ICC-02/05-01/07, 4 June 2007. The Hague: International Criminal Court.

—2007d. *Situation in the Democratic Republic of the Congo in the Case of the Prosecutor v. Thomas Lubanga Dyilo.* Appeals Chamber, Case Number ICC-01/04-01/06 OA8, 13 June 2007. The Hague: International Criminal Court.

—2010. *Kenya: Questions and Answers. Understanding the International Criminal Court.* The Hague: International Criminal Court, 31 March. ICC-PIDS-PIS-KEN-00-002/10_ENG.

—2011a. "Central African Republic. ICC-01/05-01/08. Case The Prosecutor v. Jean-Pierre Bemba Gombo." <http://www.icc-cpi.int/Menus/ICC/Situations+and+Cases/Situations/Situation+ICC+0105/Related+Cases/ICC+0105+0108/Case+The+Prosecutor+v+Jean-Pierre+Bemba+Gombo.htm> (Accessed 20 September 2011).

—2011b. "Darfur, Sudan. ICC-02/05-01/09. Case The Prosecutor v. Omar Hassan Ahmad Al Bashir." <http://www.icc-cpi.int/menus/icc/situations%20and%20cases/situations/situation%20icc%200205/related%20cases/icc02050109/icc02050109?lan=en-GB> (Accessed 20 September 2011).

—2011c. "Democratic Republic of the Congo." <http://www.icc-cpi.int/Menus/ICC/Situations+and+Cases/Situations/Situation+ICC+0104/> (Accessed 20 September 2011).

—2011d. "ICC at a Glance." <http://www.icc-cpi.int/Menus/ICC/About+the+Court/ICC+at+a+glance/> (Accessed 18 September 2011).

—2011e. "Libya. ICC-01/11. Situation in the Libyan Arab Jamahiriya." <http://www.icc-cpi.int/Menus/ICC/Situations+and+Cases/Situations/ICC0111/Situation+Index.htm> (Accessed 20 September 2011).

—2011f. "Office of the Prosecutor." <http://www.icc-cpi.int/Menus/ICC/Structure+of+the+Court/Office+of+the+Prosecutor/> (Accessed 29 September 2011).

—2011g. *Questions and Answers on the ICC Proceedings in the Libya Situation Following the Prosecutor's Request for Three Arrest Warrants*. The Hague: International Criminal Court, 16 May. ICC-PIDS-Q&A-LIB-00-002/11_ENG.

—2011h. "Situations and Cases." <http://www.icc-cpi.int/Menus/ICC/Situations+and+Cases/> (Accessed 18 September 2011).

—2012. "ICC First Verdict: Thomas Lubanga guilty of conscripting and enlisting children under the age of 15 and using them to participate in hostilities." <http://www.icc-cpi.int/NR/exeres/A70A5D27-18B4-4294-816F-BE68155242E0.htm> (Accessed 15 March 2012).

"International Criminal Court Statute Becomes Effective." 2002. *International Law Update* 8(4).

"International Justice." 2001. *The Christian Century*, 17 January.

Irwin, Rosalind. 2001. "Linking Ethics and Security in Canadian Foreign Policy." In *Ethics and Security in Canadian Foreign Policy*, ed. Rosalind Irwin. Vancouver: UBC Press, pp. 3–13.

Islami Some'a, Reza. 1994. *The Need and Prospects for an International Criminal Court*. Ph.D. diss. McGill University.

Jalloh, Charles C. 2011. "Situation in the Republic of Kenya." *The American Journal of International Law* 105(3): 540–7.

Johansen, Robert C. 1998. "Enhancing United Nations Peace-keeping." In *The Future of the United Nations System: Potential for the Twenty-first Century*, ed. Chadwick F. Alger. Tokyo: United Nations University Press, pp. 89–126.

Keating, Tom. 1993. *Canada and World Order: The Multilateralist Tradition in Canadian Foreign Policy*. Toronto: McClelland & Stewart.

Kennett, Lee, and James LaVerne Anderson. 1975. *The Gun in America: The Origins of a National Dilemma*. Westport, CT: Greenwood Press.

Kingdon, John W. 1984. *Agendas, Alternatives, and Public Policies*. Boston, MA: Little, Brown.

Kingman, Sharon. 2000. "Progress Made in Reducing the Number of Landmines Worldwide." *Bulletin of the World Health Organization* 78(11): 1370.
Kirkey, Christopher. 2001. *Washington's Response to the Ottawa Land Mines Process*. Orono, ME: The Canadian-American Center.
Kissinger, Henry A. 1957. *Nuclear Weapons and Foreign Policy*. New York: Harper.
Klare, Michael T. 1999. "The International Trade in Light Weapons: What Have We Learned?" In *Light Weapons and Civil Conflict: Controlling the Tools of Violence*, eds. Jeffrey Boutwell and Michael T. Klare. Lanham, MD: Rowman & Littlefield, pp. 9–27.
—2001. "UN Forges Program to Combat Illicit Trade in Small Arms." *Issues in Science and Technology* 18(1): 34.
Klare, Michael, and Robert I. Rotberg. 1999. *The Scourge of Small Arms*. Cambridge, MA: World Peace Foundation.
Knight, W. Andy. 2001. "Soft Power, Moral Suasion, and Establishing the International Criminal Court: Canadian Contributions." In *Ethics and Security in Canadian Foreign Policy*, ed. Rosalind Irwin. Vancouver: UBC Press, pp. 113–37.
Kolodziej, Edward A. 1992. "Renaissance in Security Studies? Caveat Lector!" *International Studies Quarterly* 36(4): 421–38.
Kongstad, Steffen. 1999. "The Continuation of the Ottawa Process: Intersessional Work and the Role of Geneva." *Disarmament Forum: Framework for a Mine-Free World* 4: 58–62.
Krasner, Stephen D. 1978. *Defending the National Interest: Raw Materials Investments and U.S. Foreign Policy*. Princeton, NJ: Princeton University Press.
—1983. "Structural Causes and Regime Consequences: Regimes as Intervening Variables." In *International Regimes*, ed. Stephen D. Krasner. Ithaca, NY: Cornell University Press, pp. 1–21.
—1999. *Sovereignty: Organized Hypocrisy*. Princeton, NJ: Princeton University Press.
Krause, Keith, and Michael C. Williams, eds. 1997. *Critical Security Studies: Concepts and Cases*. Minneapolis, MN: University of Minnesota Press.
Lamy, Steven L. 2002. "The G8 and the Human Security Agenda." In *New Directions in Global Political Governance: The G8 and International Order in the Twenty-First Century*, eds. John J. Kirton and Junichi Takase. Aldershot, UK: Ashgate, pp. 167–87.
Langille, H. Peter. 2000. "Conflict Prevention: Options for Rapid Deployment and UN Standing Forces." In *Peacekeeping and Conflict Resolution*, eds. Tom Woodhouse and Oliver Ramsbotham. London: Frank Cass, pp. 219–53.
Lapid, Yosef. 1989. "The Third Debate: On the Prospects of International Theory in a Post-Positivist Era." *International Studies Quarterly* 33: 235–54.
Latham, Andrew. 1996. *Light Weapons and International Security: A Canadian Perspective*. Toronto: York University Centre for International and Security Studies.
Lawson, Robert. 1998. "The Ottawa Process: Fast-Track Diplomacy and the International Movement to Ban Anti-Personnel Mines." In *Canada Among Nations 1998: Leadership and Dialogue*, eds. Fen Osler Hampson and Maureen Appel Molot. Don Mills, ON: Oxford University Press, pp. 81–98.
Lawson, Robert J., Mark Gwozdecky, Jill Sinclair, and Ralph Lysyshyn. 1998. "The Ottawa Process and the International Movement to Ban Anti-Personnel Mines."

In *To Walk Without Fear: The Global Movement to Ban Landmines*, eds. Maxwell A. Cameron, Robert J. Lawson, and Brian W. Tomlin. Don Mills, ON: Oxford University Press, pp. 160–84.

Leahy, Patrick J. 1997. "Toward a Global Ban on Landmines." *Technology Review* 100(7): 45.

Leguey-Feilleux, Jean-Robert. 2009. *The Dynamics of Diplomacy*. Boulder, CO: Lynne Rienner Publishers.

Leigh, Monroe. 2001. "The United States and the Statute of Rome." *American Journal of International Law* 95(1): 124–31.

Lenarcic, David A. 1998. *Knight-Errant?: Canada and the Crusade to Ban Anti-Personnel Land Mines*. Toronto: Irwin.

Levene, Abigail. 2003. "U.S. Stays Away as Global Criminal Court Gets Going." *Reuters*, 10 March. <http://story.news.yahoo.com/news?tmpl=story&u=/nm/20030311/wl_nm/dutch_warcrimes_dc_1> (Accessed 11 March 2003).

Levinson, Sanford. 1989. "The Embarrassing Second Amendment." *Yale Law Review* 99: 637–44.

Lie, Trygve. 1954. *In the Cause of Peace*. London: Macmillan.

Lobe, Jim. 2003a. "U.S. Cuts Military Aid to Friendly Nations." *OneWorld.net*, 1 October. <http://www.commondreams.org/headlines03/1001-01.htm> (Accessed 17 July 2003).

—2003b. "U.S. Punishes 35 Countries for Signing Onto International Court."*Inter Press Service*, 2 July.

Lozano, Graciela U. 1999. "The United Nations and the Control of Light Weapons." In *Light Weapons and Civil Conflict: Controlling the Tools of Violence*, eds. Jeffrey Boutwell and Michael T. Klare. Lanham, MD: Rowman & Littlefield, pp. 161–71.

Lumpe, Lora. 1999. "The Leader of the Pack." *Bulletin of the Atomic Scientists* 55(1): 27.

Lynn-Jones, Sean M. 1999. "Realism and Security Studies." In *Contemporary Security and Strategy*, ed. Craig A. Snyder. New York: Routledge, pp. 53–76.

Macdonald, Oliver A. K. 1998. "Recent Developments in Peacekeeping—The Irish Military Experience." In *A Future for Peacekeeping?*, ed. Edward Moxon-Browne. London: Macmillan, pp. 40–57.

MacFarlane, S. Neil and Yuen Foong Khong. 2006. *Human Security and the UN: A Critical History*. Bloomington, IN: Indiana University Press.

Maley, William. 2002. "The UN, NGOs, and the Land-mines Initiative: An Australian Perspective." In *Enhancing Global Governance: Towards A New Diplomacy?*, eds. Andrew F. Cooper, John English, and Ramesh Thakur. New York: United Nations University Press, pp. 90–105.

Manley, Andrew. 1998. "Anti-Landmines Crusade Wins A New Royal Champion." *The Middle East* 281(August): 50.

Maresca, Louis, and Stuart Maslen, eds. 2000. *The Banning of Anti-Personnel Landmines: The Legal Contribution of the International Committee of the Red Cross*. Cambridge, UK: Cambridge University Press.

Marx, Karl, and Frederick Engels. [1848] 1986. *Manifesto of the Communist Party*. Chicago, IL: C. H. Kerr.

Matheson, Michael J. 1997. "The Revision of the Mines Protocol." *American Journal of International Law* 91(1): 158–67.

Mathews, Jessica T. 1989. "Redefining Security." *Foreign Affairs* 68(2): 162–77.

—1997. "Power Shift." *Foreign Affairs* 76(1): 50–66.

Matthew, Richard A., and Ken R. Rutherford. 1999. "Banning Landmines in the American Century." *International Journal on World Peace* 16(2): 23.
Mayor, Federico. 1995. "Toward Global Human Security." In *People: From Impoverishment to Empowerment*, eds. Üner Kirdar and Leonard Silk. New York: New York University Press, pp. 365–9.
McElroy, John H. 1999. *American Beliefs: What Keeps a Big Country and a Diverse People United*. Chicago, IL: Ivan R. Dee.
McRae, Rob. 2001a. "Human Security, Connectivity, and the New Global Civil Society." In *Human Security and the New Diplomacy: Protecting People, Promoting Peace*, eds. Rob McRae and Don Hubert. Montreal & Kingston: McGill-Queen's University Press, pp. 236–49.
—2001b. "Human Security in a Globalized World." In *Human Security and the New Diplomacy: Protecting People, Promoting Peace*, eds. Rob McRae and Don Hubert. Montreal & Kingston: McGill-Queen's University Press, pp. 14–27.
Meek, Sarah. 2000. "Combating Arms Trafficking: Progress and Prospects." In *Running Guns: The Global Black Market in Small Arms*, ed. Lora Lumpe. London: Zed Books, pp. 183–206.
Meron, Theodor. 1999. "Crimes Under the Jurisdiction of the International Criminal Court." In *Reflections on the International Criminal Court: Essays in Honour of Adriaan Bos*, eds. Herman A. M. von Hebel, Johan G. Lammers, and Jolien Schukking. The Hague: T. M. C. Asser, pp. 47–55.
Meyer, Karl E. 2002. "Criminal Thinking in Washington." *World Policy Journal* 19(2): 106–8.
Milner, Helen. 1987. "Resisting the Protectionist Temptation: Industry and the Making of Trade Policy in France and the United States During the 1970s." *International Organization* 41(4): 639–65.
Morgan, Patrick. 2007. "Security in International Politics: Traditional Approaches." In *Contemporary Security Studies*, ed. Alan Collins. New York: Oxford University Press, pp. 13–34.
Morgenthau, Hans J. 1948. *Politics Among Nations: The Struggle for Power and Peace*. New York: Alfred A. Knopf.
—1978. *Politics Among Nations: The Struggle for Power and Peace*. Fifth Edition, Revised. New York: Alfred A. Knopf.
Morrison, Philip, and Kosta Tsipis. 1997. "New Hope in the Minefields." *MIT's Technology Review* 100(7): 38–46.
Morton, Jeffrey S. 2000. *The International Law Commission of the United Nations*. Columbia, SC: University of South Carolina Press.
Multinational Standby High Readiness Brigade for United Nations Operations. 2001a. "SHIRBRIG History." <http://www.shirbrig.dk> (Accessed 15 August 2002).
—2001b. "SHIRBRIG Organisation." <http://www.shirbrig.dk> (Accessed 11 January 2003).
—2001c. "Standard Briefing." November. <http://www.shirbrig.dk> (Accessed 11 January 2003).
—2006. "Deployment to Sudan." <http://www.shirbrig.dk/shirbrig/html/Sudandepl.htm> (Accessed 24 January 2007).
—2007a. "UNMEE." <http://www.shirbrig.dk/shirbrig/html/unmee.htm> (Accessed 24 January 2007).
—2007b. "UNMIL." <http://www.shirbrig.dk/shirbrig/html/unmil.htm> (Accessed 24 January 2007).

—2010a. "Facts." <http://www.shirbrig.dk/html/facts.htm> (Accessed 21 September 2010).
—2010b. "Introduction to SHIRBRIG." <http://www.shirbrig.dk/html/sb_intro.htm> (Accessed 10 July 2010).
Murphy, John F. 2010. "Gulliver No Longer Quivers: U.S. Views on and the Future of the International Criminal Court." *International Lawyer* 44(4): 1123.
Mutimer, David. 1999. "Beyond Strategy: Critical Thinking and the New Security Studies." In *Contemporary Security and Strategy*, ed. Craig A. Snyder. New York: Routledge, pp. 77–101.
—2007. "Critical Security Studies: A Schismatic History." In *Contemporary Security Studies*, ed. Alan Collins. New York: Oxford University Press, pp. 53–74.
Nash, William L. 2000. "The ICC and the Deployment of U.S. Armed Forces." In *The United States and the International Criminal Court: National Security and International Law*, eds. Sarah B. Sewall and Carl Kaysen. Lanham, MD: Rowman & Littlefield, pp. 153–64.
National Wildlife Federation. 2000. "Land Mines: Grim Legacy for Wildlife Worldwide." *International Wildlife*, March–April.
Neack, Laura. 2000. "Middle Powers Once Removed: The Diminished Global Role of Middle Powers and American Grand Strategy." Paper presented at the International Studies Association 41st Annual Convention, Los Angeles, CA, March 14–18, 2000. Available from *Columbia International Affairs Online.*
The Netherlands, Non-paper. 1995. *A UN Rapid Deployment Brigade: 'A Preliminary Study.'* Revised Version, April 1995. In *Letter dated 7 April 1995 from the Permanent Representative of the Netherlands to the United Nations Addressed to the Secretary-General.* UN General Assembly and Security Council, A/49/886-S/1995/276, 10 April.
The Netherlands. Permanent Mission of the Kingdom of the Netherlands to the United Nations. 2003. *National Statement, UN Biennial Meeting on the Implementation of the UN Program of Action on Small Arms and Light Weapons in All Its Aspects, by H. E. Mr. Chris Sanders, Ambassador to the Conference of Disarmament of The Netherlands in Geneva.* New York, 7 July.
Neumann, Iver B. 2002. "Harnessing Social Power: State Diplomacy and the Landmines Issue." In *Enhancing Global Governance: Towards A New Diplomacy?*, eds. Andrew F. Cooper, John English, and Ramesh Thakur. New York: United Nations University Press, pp. 106–32.
NGO Committee on Disarmament. 2001a. "NGOs Played Major Role." *Disarmament Times* 24(Summer, Special Issue on Small Arms): 3.
—2001b. "Significance of the Program of Action." *Disarmament Times* 24(Summer, Special Issue on Small Arms): 2.
—2001c. "Small Arms Conference Sets Significant Action But Key Provisions Are Weak." *Disarmament Times* 24(Summer, Special Issue on Small Arms): 1–6.
—2001d. "UN-Small Arms Time Line." *Disarmament Times* 24(Summer, Special Issue on Small Arms): 1–8.
Nigeria. Permanent Mission of Nigeria to the United Nations. 2003. *Statement by Chuku Udedibia, Minister, at the First Biennial Meeting of States to Consider the Implementation of the Programme of Action to Prevent, Combat and Eradicate the Illicit Trade in Small Arms and Light Weapons in All Its Aspects.* New York, 7 July.

—2005. *Statement by Mr. Chuku Udedibia, Minister, Permanent Mission of Nigeria to the UN, On Behalf of the African Group, Delivered at the Second Biennial Meeting of States to Consider the Implementation of the Programme of Action to Prevent, Combat and Eradicate the Illicit Trade in Small Arms and Light Weapons in All Its Aspects*. New York, 11 July.

Norway. Permanent Mission of Norway to the United Nations. 2003. *First Biennial Meeting of States to Consider the Implementation of the Programme of Action to Prevent, Combat and Eradicate the Illicit Trade in Small Arms and Light Weapons in All Its Aspects, Statement by H. E. Mr. Ole Peter Kolby, Ambassador, Permanent Representative*. New York, 7 July.

—2005. *Statement by H. E. Ms. Mona Juul, Chargé d'Affaires, at the Second Biennial Meeting of States to Consider the Implementation of the UN Programme of Action to Prevent, Combat and Eradicate the Illicit Trade in Small Arms and Light Weapons in All Its Aspects*. New York, 11 July.

Nuechterlein, Donald E. 2001. *America Recommitted: A Superpower Assesses Its Role in a Turbulent World*. Second Edition. Lexington, KY: The University Press of Kentucky.

Nye, Joseph S., Jr. 1990. *Bound to Lead: The Changing Nature of American Power*. New York: Basic Books.

Obama, Barack. 2007. "Renewing American Leadership." *Foreign Affairs* 86(4): 2–16.

O'Connor, Karen and Graham Barron. 1998. "Madison's Mistake? Judicial Construction of the Second Amendment." In *The Changing Politics of Gun Control*, eds. John M. Bruce and Clyde Wilcox. Lanham, MD: Rowman & Littlefield, pp. 74–87.

Omestad, Thomas. 1998. "U.S. Seeks to Weaken Court." *U.S. News & World Report*, 29 June.

Organization of American States. 2011. "Inter-American Convention Against the Illicit Manufacturing of and Trafficking in Firearms, Ammunition, Explosives and Other Related Materials." <http://www.oas.org/juridico/english/sigs/a-63.html> (Accessed 21 September 2011).

The Oslo Meeting on Small Arms. 1998. "An International Agenda on Small Arms and Light Weapons: Elements of a Common Understanding." 13–14 July. <http://www.iansa.org/documents/regional/2000/jan_00/oslomeeting.htm> (Accessed 26 April 2003).

Owen, Taylor. 2008. "The Uncertain Future of Human Security in the UN." In *Rethinking Human Security*, eds. Moufida Goucha and John Crowley. Malden, MA: Wiley-Blackwell & UNESCO, pp. 113–27.

Pace, William R. 1999. "The Relationship Between the International Criminal Court and Non-Governmental Organizations." In *Reflections on the International Criminal Court: Essays in Honour of Adriaan Bos*, eds. Herman A. M. von Hebel, Johan G. Lammers, and Jolien Schukking. The Hague: T. M. C. Asser, pp. 189–211.

Pace, William R., and Jennifer Schense. 2001. "Coalition for the International Criminal Court at the Preparatory Commission." In *The International Criminal Court: Elements of Crimes and Rules of Procedure and Evidence*, ed. Roy S. Lee. Ardsley, NY: Transnational, pp. 705–34.

Page, Edward, and Michael Redclift. 2002. "Introduction: Human Security and the Environment at the New Millennium." In *Human Security and the*

Environment: International Comparisons, eds. Edward A. Page and Michael Redclift. Cheltenham, UK: Edward Elgar, pp. 1–24.

Painchaud, Paul. 1965. "Middlepowermanship as an Ideology." In *Canada's Role as a Middle Power; Papers Given at the Third Annual Banff Conference on World Development, August 1965*, ed. J. King Gordon. Toronto: Canadian Institute of International Affairs, pp. 29–35.

Panel on United Nations Peace Operations. 2000. *Report of the Panel on United Nations Peace Operations*. UN General Assembly and Security Council, A/55/305-S/2000/809, 21 August.

Paris, Roland. 2001. "Human Security: Paradigm Shift or Hot Air?" *International Security* 26(2): 87–102.

Parker, Sarah. 2010. *National Implementation of the United Nations Small Arms Programme of Action and the International Tracing Instrument: An Analysis of Reporting in 2009–10*. Geneva: Small Arms Survey, Graduate Institute of International and Development Studies, June.

"Permanent International Criminal Court Established." 1998. *UN Chronicle* 35(3): 10.

Pettiford, Lloyd, and Melissa Curley. 1999. *Changing Security Agendas and the Third World*. New York: Pinter.

Pfaff, William. 1998. "Faulty Judgment." *Commonweal* 125(15): 9–10.

Pirnie, Bruce R., and William E. Simons. 1996. *Soldiers For Peace: An Operational Typology*. Santa Monica, CA: RAND.

Pisik, Betsy. 1998. "World Court Goes to Trial." *Insight on the News* 14(45): 41.

Pratt, Cranford, ed. 1989. *Internationalism Under Strain: The North-South Policies of Canada, the Netherlands, Norway, and Sweden*. Toronto: University of Toronto Press.

—1990. *Middle Power Internationalism: The North-South Dimension*. Kingston & Montreal: McGill-Queen's University Press.

Price, Richard. 1998. "Reversing the Gun Sights: Transnational Civil Society Targets Land Mines." *International Organization* 52(3): 613–44.

Purves, Bill. 2001. *Living with Landmines: From International Treaty to Reality*. Montreal: Black Rose Books.

Ratner, Steven. 1995. "Peacemaking, Peacekeeping, and Peace Enforcement: Conceptual and Legal Underpinnings of the U.N. Role." In *U.N. Peacekeeping: Japanese and American Perspectives*, eds. Selig S. Harrison and Masashi Nishihara. Washington, DC: Carnegie Endowment for International Peace, pp. 17–30.

Renner, Michael. 1999. "Arms Control Orphans." *Bulletin of the Atomic Scientists* 55(1): 22.

Roach, Steven C. 2006. *Politicizing the International Criminal Court: The Convergence of Politics, Ethics, and Law*. Lanham, MD: Rowman & Littlefield.

Roberts, Shawn, and Jody Williams. 1995. *After The Guns Fall Silent: The Enduring Legacy of Landmines*. First Edition. Washington, DC: Vietnam Veterans of America Foundation.

Robinson, Darryl. 2001. "Case Study: The International Criminal Court." In *Human Security and the New Diplomacy: Protecting People, Promoting Peace*, eds. Rob McRae and Don Hubert. Montreal & Kingston: McGill-Queen's University Press, pp. 170–77.

Robinson, Darryl, and Valerie Oosterveld. 2001. "The Evolution of International Humanitarian Law." In *Human Security and the New Diplomacy: Protecting People, Promoting Peace*, eds. Rob McRae and Don Hubert. Montreal & Kingston: McGill-Queen's University Press, pp. 161–9.

Rogers, Paul. 2007. "Peace Studies." In *Contemporary Security Studies*, ed. Alan Collins. New York: Oxford University Press, pp. 35–52.

The Rome Statute of the International Criminal Court: A Commentary. Materials. 2002. Oxford: Oxford University Press.

Roth, Kenneth. 1998. "Sidelined on Human Rights: America Bows Out." *Foreign Affairs* 77(2): 2–6.

—2009. "Was the Iraq War a Humanitarian Intervention? And What Are Our Responsibilities Today?" In *Responsibility to Protect: The Global Moral Compact for the 21st Century*, eds. Richard H. Cooper and Juliette Voïnov Kohler. New York: Palgrave Macmillan, pp. 101–13.

Sadat, Leila N. 2000. "The Evolution of the ICC: From The Hague to Rome and Back Again." In *The United States and the International Criminal Court: National Security and International Law*, eds. Sarah B. Sewall and Carl Kaysen. Lanham, MD: Rowman & Littlefield, pp. 31–49.

Sage, Colin. 2002. "Food Security." In *Human Security and the Environment: International Comparisons*, eds. Edward A. Page and Michael Redclift. Cheltenham, UK: Edward Elgar, pp. 128–53.

Schabas, William A. 2011. *An Introduction to the International Criminal Court*. Fourth Edition. Cambridge: Cambridge University Press.

Scheffer, David J. 1999. "The United States and the International Criminal Court." *American Journal of International Law* 93(1): 12–22.

Schelling, Thomas C. 1966. *Arms and Influence*. New Haven, CT: Yale University Press.

"Sign On, Opt Out; What? A World Criminal Court!" 2001. *The Economist*, 6 January.

Sikkink, Kathryn. 2002. "Restructuring World Politics: The Limits and Asymmetries of Soft Power." In *Restructuring World Politics: Transnational Social Movements, Networks, and Norms*, eds. Sanjeev Khagram, James V. Riker, and Kathryn Sikkink. Minneapolis, MN: University of Minnesota Press, pp. 301–17.

Smaldone, Joseph P. 1999. "Mali and the West African Light Weapons Moratorium." In *Light Weapons and Civil Conflict: Controlling the Tools of Violence*, eds. Jeffrey Boutwell and Michael T. Klare. Lanham, MD: Rowman & Littlefield, pp. 129–45.

Small, Michael. 2001. "Case Study: The Human Security Network." In *Human Security and the New Diplomacy: Protecting People, Promoting Peace*, eds. Rob McRae and Don Hubert. Montreal & Kingston: McGill-Queen's University Press, pp. 231–5.

Snyder, Jack. 1991. *Myths of Empire: Domestic Politics and International Ambition*. Ithaca, NY: Cornell University Press.

Spalding, Matthew, ed. 2002. *The Founders' Almanac: A Practical Guide to the Notable Events, Greatest Leaders & Most Eloquent Words of the American Founding*. Reference Edition. Washington, DC: The Heritage Foundation.

Spitzer, Robert J. 1995. *The Politics of Gun Control*. Chatham, NJ: Chatham House.

Stahn, Carsten. 2007. "Responsibility to Protect: Political Rhetoric or Emerging Legal Norm?" *The American Journal of International Law* 101(1): 99–120.
Stoett, Peter. 1999. *Human and Global Security: An Exploration of Terms*. Toronto: University of Toronto Press.
Stokke, Olav, ed. 1989. *Western Middle Powers and Global Poverty: The Determinants of the Aid Policies of Canada, Denmark, the Netherlands, Norway, and Sweden*. Uppsala, Sweden: The Scandinavian Institute of African Studies.
Stripple, Johannes. 2002. "Climate Change as a Security Issue." In *Human Security and the Environment: International Comparisons*, eds. Edward A. Page and Michael Redclift. Cheltenham, UK: Edward Elgar, pp. 105–27.
Szasz, Paul. 1980. "The Conference on Excessively Injurious or Indiscriminate Weapons." *American Journal of International Law* 74(1): 212–15.
Taylor, Paul, Sam Daws, and Ute Adamczick-Gerteis, eds. 1997. "Key Elements of the Clinton Administration's Policy on Reforming Multilateral Peace Operations." In *Documents on Reform of the United Nations*, Aldershot, UK: Dartmouth, pp. 123–37.
Tepperman, Jonathan D. 2000. "Contempt of Court." *Washington Monthly* 32(11): 25.
Trubowitz, Peter. 1998. *Defining the National Interest: Conflict and Change in American Foreign Policy*. Chicago, IL: University of Chicago Press.
Turner, Mandy, Neil Cooper, and Michael Pugh. 2011. "Institutionalised and Co-opted: Why Human Security Has Lost its Way." In *Critical Perspectives on Human Security: Rethinking Emancipation and Power in International Relations*, eds. David Chandler and Nik Hynek. New York: Routledge, pp. 83–96.
Turner, Terisa E., Leigh S. Brownhill, and Wahu M. Kaara. 2001. "Gender, Food Security, and Foreign Policy Towards Africa: Women Farmers in Kenya and the Right to Sustenance." In *Ethics and Security in Canadian Foreign Policy*, ed. Rosalind Irwin. Vancouver: UBC Press, pp. 95–110.
Ul Haq, Mahbub. 1995. "New Imperatives of Human Security." In *People: From Impoverishment to Empowerment*, eds. Üner Kirdar and Leonard Silk. New York: New York University Press, pp. 370–9.
"UN Conference on Illicit Trade in Small Arms." 2001. *American Journal of International Law* 95(4): 901–3.
"Under the Gun." 2001. *UN Chronicle* 38(1): 13–14.
United Kingdom. 2005. *Biennial Meeting of States to Consider the Implementation of the UN Programme of Action on Small Arms and Light Weapons, Statement by H. E. Ambassador John Freeman, Head of UK Delegation to the Biennial Meeting of States, on Behalf of the European Union*. New York, 11 July.
United Nations. 2001. *Report of the United Nations Conference on the Illicit Trade in Small Arms and Light Weapons in All Its Aspects: New York, 9–20 July 2001*. UN General Assembly, A/CONF.192/15.
—2004. *A More Secure World: Our Shared Responsibility*. Report of the Secretary-General's High-level Panel on Threats, Challenges and Change. New York: United Nations Department of Public Information.
—2011a. "Charter of the United Nations. Chapter 1: Purposes and Principles." <http:// www. un.org/en/documents/charter/chapter1. shtml> (Accessed 7 September 2011).

—2011b. "United Nations Peacekeeping." <http://www.un.org/en/peacekeeping/> (Accessed 12 September 2011).

United Nations Assistance Mission in Afghanistan. 2011. "UNAMA: United Nations Assistance Mission in Afghanistan." <http://unama.unmissions.org/Default.aspx?tabid=1741> (Accessed 12 September 2011).

United Nations Conference to Review Progress Made in the Implementation of the Programme of Action to Prevent, Combat and Eradicate the Illicit Trade in Small Arms and Light Weapons in All Its Aspects. 2006a. *Developing Common Guidelines for National Controls on Transfers of Small Arms and Light Weapons: Progress Since 2003*. Working paper submitted by the United Kingdom of Great Britain and Northern Ireland. UN General Assembly, A/CONF.192/2006/RC/WP.1, 19 June.

—2006b. *Programme of Action to Prevent, Combat and Eradicate the Illicit Trade in Small Arms and Light Weapons in All Its Aspects: A Strategy for Further Implementation*. Working Paper submitted by the President. UN General Assembly, A/CONF.192/2006/RC/WP.4, 29 June.

United Nations Department for Disarmament Affairs. 2003. "First Biennial Meeting on the Implementation of the Programme of Action to Prevent, Combat and Eradicate the Illicit Trade in Small Arms and Light Weapons In All Its Aspects." <http://disarmament.un.org/cab/salw-2003.html> (Accessed 7 August 2003).

United Nations Department of Public Information. Peace and Security Section. 2005. "Ethiopia and Eritrea—UNMEE—Background." <http://www.un.org/Depts/dpko/missions/unmee/background.html> (Accessed 20 January 2007).

—2007. *The United Nations and Darfur. Fact Sheet*. <http://www.un.org/News/dh/infocus/sudan/fact_sheet.pdf> (Accessed 21 July 2007).

United Nations Development Program. 1994. *Human Development Report*. New York: Oxford University Press.

United Nations General Assembly. 2001. *Protocol against the Illicit Manufacturing of and Trafficking in Firearms, Their Parts and Components and Ammunition, Supplementing the United Nations Convention against Transnational Organized Crime*. UN General Assembly Resolution, A/RES/55/255, 8 June. <http://www.unodc.org/unodc/crime_cicp_signatures_firearms.html> (Accessed 25 July 2007).

—2005a. *In Larger Freedom: Towards Development, Security and Human Rights for All*. Report of the Secretary-General, A/59/2005, 21 March.

—2005b. *Report of the Second Biennial Meeting of States to Consider the Implementation of the Programme of Action to Prevent, Combat and Eradicate the Illicit Trade in Small Arms and Light Weapons in All Its Aspects*. UN General Assembly, A/CONF.192/BMS/2005/1, 19 July.

—2005c. *2005 World Summit Outcome*. Draft resolution referred to the High-level Plenary Meeting of the General Assembly by the General Assembly at its fifty-ninth session, A/60/L.1, 15 September.

—2006. *Report of the United Nations Conference to Review Progress Made in the Implementation of the Programme of Action to Prevent, Combat and Eradicate the Illicit Trade in Small Arms and Light Weapons in All Its Aspects*. UN General Assembly, A/CONF.192/2006/RC/9, 12 July.

—2008. *Report of the Third Biennial Meeting of States to Consider the Implementation of the Programme of Action to Prevent, Combat and Eradicate the Illicit Trade in Small Arms and Light Weapons in All Its Aspects*. UN General Assembly, A/CONF.192/BMS/2008/3, 20 August.

—2009. *Implementing the Responsibility to Protect*. Report of the Secretary-General, A/63/677, 12 January.

—2010. *Report of the Fourth Biennial Meeting of States to Consider the Implementation of the Programme of Action to Prevent, Combat and Eradicate the Illicit Trade in Small Arms and Light Weapons in All Its Aspects*. UN General Assembly, A/CONF.192/BMS/2010/3, 30 June.

United Nations International Study on Firearm Regulation. 1998. New York: United Nations.

United Nations Military Adviser. 2002. "Annual Update on the United Nations Standby Arrangements System." Presentation by the acting Military Adviser for the Annual Update on Military Aspects of the United Nations Stand-by Arrangements System, 12 December. <http://www.un.org/Depts/dpko/rapid/AnnualUpdate.html> (Accessed 19 January 2007).

United Nations Military Division. Department of Peacekeeping Operations. 2001. "UN Standby Arrangements System Military Handbook." <http://www.un.org/Depts/dpko/rapid/Handbook.html> (Accessed 15 August 2002).

—2003a. "Missions Planned with the Support of the United Nations Standby Arrangements System." <http://www.un.org/Depts/dpko/rapid/Missions.html> (Accessed 19 January 2007).

—2003b. "United Nations Stand-by Arrangements System. Background." <http://www.un.org/Depts/dpko/rapid/Background.html> (Accessed 19 January 2007).

United Nations Mission in Kosovo. 2011. "About UNMIK." <http://www.unmikonline.org/Pages/about.aspx> (Accessed 12 September 2011).

United Nations News Service. 2007. "UN Team on Peacekeeping Options Heads to Central African Republic in Next Two Weeks." New York: Office of the Spokesman for the Secretary-General, 15 January. <http://www.globalsecurity.org/military/library/news/2007/01/mil-070115-unnews01.htm> (Accessed 19 February 2007).

United Nations Office at Geneva. 2011. "Disarmament: States Parties and Signatories CCW." <http://www.unog.ch/__80256ee600585943.nsf/%28httpPages%29/3ce7cfc0aa4a7548c12571c00039cb0c?OpenDocument&ExpandSection=1#_Section1> (Accessed 21 November 2011).

United Nations Office for Disarmament Affairs. 2011. "Arms Trade Treaty." <http://www.un.org/disarmament/convarms/ArmsTradeTreaty/html/ATTMeetings2009-11.shtml> (Accessed 9 October 2011).

United Nations Preparatory Committee for the United Nations Conference on the Illicit Trade in Small Arms and Light Weapons in All Its Aspects. 2000. *Working Paper Submitted by Canada Entitled "Towards a United Nations 2001 Action Plan on Small Arms."* UN General Assembly, A/CONF.192/PC/14, 21 July.

United Nations Preparatory Committee for the United Nations Conference to Review Progress Made in the Implementation of the Programme of Action to Prevent, Combat and Eradicate the Illicit Trade in Small Arms and Light Weapons in All Its Aspects, New York, 9–20 January 2006. 2006a. *Importance of the Subject on Civilian Possession in the Combat Against the Illicit Trade of Small Arms and Light Weapons*. Concept paper submitted by Mexico. UN General Assembly, A/CONF.192/2006/PC/CRP.7, 9 January.

—2006b. *Preparing for the 2006 SALW PoA Review Conference: Addressing the Demand for Illicit Small Arms and Light Weapons*. Paper submitted by Canada. UN General Assembly, A/CONF.192/2006/PC/CRP.15, 18 January.

—2006c. *The Strengthening of Controls over Transfers (Import, Export and Transit) of Small Arms and Light Weapons*. Paper submitted by Brazil. UN General Assembly, A/CONF.192/2006/PC/CRP.11, 17 January.

United Nations Program of Action Implementation Support System. 2011. "Firearms Protocol." <http://www.poa-iss.org/FirearmsProtocol/FirearmsProtocol.aspx> (Accessed 21 September 2011).

United Nations Secretary-General. 1992. *An Agenda for Peace: Preventive Diplomacy, Peacemaking, and Peace-Keeping*. Report of the Secretary-General pursuant to the statement adopted by the Summit Meeting of the Security Council on 31 January 1992. UN General Assembly and Security Council, A/47/277-S/24111, 17 June.

—1994a. *Improving the Capacity of the United Nations for Peace-Keeping: Report of the Secretary-General*. UN General Assembly and Security Council, A/48/403-S/26450, 14 March. Reprinted in *Documents on Reform of the United Nations*, eds. Paul Taylor, Sam Daws, and Ute Adamczick-Gerteis. Aldershot, UK: Dartmouth, 1997, pp. 67–87.

—1994b. *Stand-by Arrangements for Peace-Keeping: Report of the Secretary-General*. UN Security Council, S/1994/777, 30 June. Reprinted in *Documents on Reform of the United Nations*, eds. Paul Taylor, Sam Daws, and Ute Adamczick-Gerteis. Aldershot, UK: Dartmouth, 1997, pp. 151–2.

—1995. *Supplement to An Agenda for Peace: Position Paper of the Secretary-General on the Occasion of the Fiftieth Anniversary of the United Nations*. Report of the Secretary-General on the Work of the Organization. UN General Assembly and Security Council, A/50/60-S/1995/1, 3 January. Reprinted in *Documents on Reform of the United Nations*, eds. Paul Taylor, Sam Daws, and Ute Adamczick-Gerteis. Aldershot, UK: Dartmouth, 1997, pp. 89–115.

—1999. *Convening of an International Conference on the Illicit Arms Trade in All Its Aspects: Report of the Secretary-General*. UN General Assembly, A/54/260, 20 August.

—2000a. *Convening of an International Conference on the Illicit Arms Trade in All Its Aspects: Report of the Secretary-General*. UN General Assembly, A/54/260/Add.1, 24 February.

—2000b. *Report of the Secretary-General on Ethiopia and Eritrea*. UN Security Council, S/2000/643, 30 June.

—2000c. *Ethiopia and Eritrea: Report of the Secretary-General*. UN Security Council, S/2000/785, 9 August.

—2005. "Secretary-General Welcomes Adoption of Security Council Resolution Referring Situation in Darfur, Sudan to International Criminal Court Prosecutor." UN Secretary-General, SG/SM/9797, AFR/1132, 31 March. <http://www.un.org/News/Press/docs/2005/sgsm9797.doc.htm> (Accessed 3 February 2007).

United Nations Security Council 2000a. *On Establishment of the UN Mission in Ethiopia and Eritrea*. UN Security Council Resolution, S/RES/1312, 31 July.

—2000b. *On Deployment of Troops and Military Observers Within the UN Mission in Ethiopia and Eritrea (UNMEE)*. UN Security Council Resolution, S/RES/1320, 15 September.

—2011a. *Resolution 1970*. UN Security Council Resolution, S/RES/1970, 26 February.

—2011b. *Resolution 1996*. UN Security Council Resolution, S/RES/1996, 8 July.

United Nations Treaty Collection. 2011. "Protocol against the Illicit Manufacturing of and Trafficking in Firearms, Their Parts and Components and Ammunition, supplementing the United Nations Convention against Transnational Organized Crime." <http://treaties.un.org/Pages/ViewDetails.aspx?src=TREATY&mtdsg_no=XVIII-12-c&chapter=18&lang=en> (Accessed 21 September 2011).

United States Congress. 2010. *A Concurrent Resolution Recognizing the United States National Interest in Helping to Prevent and Mitigate Acts of Genocide and Other Mass Atrocities against Civilians, and Supporting and Encouraging Efforts to Develop a Whole of Government Approach to Prevent and Mitigate Such Acts*. 111th Cong., 2nd sess. S. Con. Res. 71.

United States Defense Intelligence Agency and United States Army Foreign Science and Technology Center. 1992. *Landmine Warfare – Trends & Projections*. December. DST-116OS-019-92.

United States Department of State. 1993. *Hidden Killers: The Global Problem with Uncleared Landmines*. Washington: U.S. Department of State, Political-Military Affairs Bureau, Office of International Security Operations, July.

—1998. "Small Arms Issues: U.S. Policy and Views." 11 August. <http://www.state.gov> (Accessed 7 August 2002).

United States Department of State. Bureau of International Organization Affairs. 2001. "UN Mission in Ethiopia and Eritrea (UNMEE)." 12 April. <http://www.state.gov/p/io/rls/fs/2001/2517.htm> (Accessed 15 August 2002).

United States Department of State. Office of the Spokesman. 2000. "The Brahimi Report on UN Peacekeeping Reform." 23 August. <http://www.state.gov/www/issues/fs-peacekp_reform_000823.html> (Accessed 27 January 2007).

Urquhart, Brian. 1993. "For a UN Volunteer Military Force." *The New York Review of Books*, 10 June.

Urquhart, Brian, and François Heisbourg. 1998. "Prospects for a Rapid Response Capability: A Dialogue." In *Peacemaking and Peacekeeping for the New Century*, eds. Olara A. Otunnu and Michael W. Doyle. Lanham, MD: Rowman & Littlefield, pp. 189–99.

Urquhart, Sir Brian, et al. 1997. "For a UN Volunteer Military Force and Four Replies." In *Documents on Reform of the United Nations*, eds. Paul Taylor, Sam Daws, and Ute Adamczick-Gerteis. Aldershot, UK: Dartmouth, pp. 139–50.

"U.S. Signing of the Statute of the International Criminal Court." 2001. *American Journal of International Law* 95(2): 397–400.

Uslaner, Eric M. 1998. "All in the Family? Interest Groups and Foreign Policy." In *Interest Group Politics*, Fifth Edition, eds. Allan J. Cigler and Burdett A. Loomis. Washington, DC: Congressional Quarterly Press, pp. 365–86.

Varner, Bill. 2009. "U.S. Backs Arms Trade Treaty at UN, Abandoning Bush Opposition." *Bloomberg*, 30 October. <http://www.bloomberg.com/apps/news?pid=newsarchive&sid=abkyS4.975YM> (Accessed 11 October 2011).

Vines, Alex. 1998. "The Crisis of Anti-Personnel Mines." In *To Walk Without Fear: The Global Movement to Ban Landmines*, eds. Maxwell A. Cameron, Robert J. Lawson, and Brian W. Tomlin. Don Mills, ON: Oxford University Press, pp. 118–35.

Von Hebel, Herman. 1999. "An International Criminal Court—A Historical Perspective." In *Reflections on the International Criminal Court: Essays in*

Honour of Adriaan Bos, eds. Herman A. M. von Hebel, Johan G. Lammers, and Jolien Schukking. The Hague: T. M. C. Asser, pp. 13–38.
Von Vorys, Karl. 1990. *American National Interest: Virtue and Power in Foreign Policy*. New York: Praeger.
Wæver, Ole. 1996. "The Rise and Fall of the Inter-Paradigm Debate." In *International Theory: Positivism and Beyond*, eds. Steve Smith, Ken Booth, and Marysia Zalewski. New York: Cambridge University Press, pp. 149–85.
Walt, Steven M. 1987. *The Origins of Alliances*. Ithaca, NY: Cornell University Press.
—1991. "The Renaissance of Security Studies." *International Studies Quarterly* 35(2): 211–39.
Waltz, Kenneth N. 1979. *Theory of International Politics*. New York: Random House.
Wareham, Mary. 1998. "Rhetoric and Policy Realities in the United States." In *To Walk Without Fear: The Global Movement to Ban Landmines*, eds. Maxwell A. Cameron, Robert J. Lawson, and Brian W. Tomlin. Don Mills, ON: Oxford University Press, pp. 212–43.
Wassenaar Arrangement on Export Controls for Conventional Arms and Dual-Use Goods and Technologies. 2011. "Introduction." <http://www.wassenaar.org/introduction/index.html> (Accessed 26 November 2011).
Weir, William. 1997. *A Well Regulated Militia: The Battle Over Gun Control*. North Haven, CT: Archon Books.
Weiss, Thomas G. 2009. "Toward a Third Generation of International Institutions: Obama's UN Policy." *The Washington Quarterly* 32(3): 141–62.
—2011. "Whither R2P?" *E-International Relations*. 31 August. <http://www.e-ir.info/?p=13421> (Accessed 8 September 2011).
Weiss, Thomas G., David P. Forsythe, and Roger A. Coate. 1997. *The United Nations and Changing World Politics*. Second Edition. Boulder, CO: Westview.
Weschler, Lawrence. 2000. "Exceptional Cases in Rome: The United States and the Struggle for an ICC." In *The United States and the International Criminal Court: National Security and International Law*, eds. Sarah B. Sewall and Carl Kaysen. Lanham, MD: Rowman & Littlefield, pp. 85–111.
Williams, Jody, and Stephen Goose. 1998. "The International Campaign to Ban Landmines." In *To Walk Without Fear: The Global Movement to Ban Landmines*, eds. Maxwell A. Cameron, Robert J. Lawson, and Brian W. Tomlin. Don Mills, ON: Oxford University Press, pp. 20–47.
Williamson, Richard L., Jr. 2000. "International Regulation of Land Mines." In *Commitment and Compliance: The Role of Non-Binding Norms in the International Legal System*, eds. Dinah Shelton. Oxford: Oxford University Press, pp. 505–21.
Winslow, Philip C. 1997. *Sowing the Dragon's Teeth: Land Mines and the Global Legacy of War*. Boston, MA: Beacon.
Wixley, Sue. 2004. "Take a Stand on New U.S. Landmine Policy." <http://www.icbl.org/news/2004/466.php> (Accessed 17 May 2004).
Wood, Bernard. 1990. "Towards North-South Middle Power Coalitions." In *Middle Power Internationalism: The North-South Dimension*, ed. Cranford Pratt. Kingston & Montreal: McGill-Queen's University Press, pp. 69–107.
World Commission on Environment and Development. 1987. *Our Common Future*. Oxford: Oxford University Press.
"World Watch: Ouagadougou." 1998. *Time International*, 28 September.

INDEX